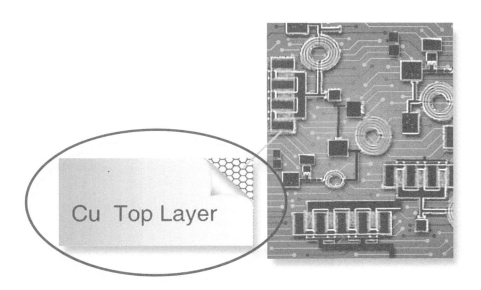

Graphene and VLSI Interconnects

Graphene and VLSI Interconnects

Cher-Ming Tan
Udit Narula
Vivek Sangwan

Published by

Jenny Stanford Publishing Pte. Ltd.
Level 34, Centennial Tower
3 Temasek Avenue
Singapore 039190

Email: editorial@jennystanford.com
Web: www.jennystanford.com

British Library Cataloguing-in-Publication Data
A catalogue record for this book is available from the British Library.

Graphene and VLSI Interconnects

Copyright © 2022 by Jenny Stanford Publishing Pte. Ltd.

All rights reserved. This book, or parts thereof, may not be reproduced in any form or by any means, electronic or mechanical, including photocopying, recording or any information storage and retrieval system now known or to be invented, without written permission from the publisher.

For photocopying of material in this volume, please pay a copying fee through the Copyright Clearance Center, Inc., 222 Rosewood Drive, Danvers, MA 01923, USA. In this case permission to photocopy is not required from the publisher.

ISBN 978-981-4877-82-4 (Hardcover)
ISBN 978-1-003-22488-4 (eBook)

Contents

Preface		ix
1. Introduction		**1**
1.1	Importance of Interconnections in VLSI	1
1.2	Copper Is Running Out of Steam	2
1.3	Graphene as a Solution	4
1.4	Properties and Applications of Graphene	9
1.5	Summary	11
2. Graphene Synthesis Methods		**17**
2.1	Overview of Existing Methods	17
	2.1.1 Mechanical Exfoliation	17
	2.1.2 Graphite Sonication	18
	2.1.3 Graphene Oxide Reduction	18
	2.1.4 Epitaxial Growth	19
2.2	Drawbacks of Current Methods with Respect to Applications in VLSI Interconnect Fabrication	20
2.3	Summary	22
3. Novel and Improved Graphene Synthesis Method		**27**
3.1	Exploration of Carbon Sources	27
3.2	Amorphous Carbon as a Promising Candidate: A PVD-Based Synthesis	28
	3.2.1 Exploration of Different Placements of Carbon Source	28
	3.2.1.1 Sample preparation using PVD method	28
	3.2.1.2 Post-PVD annealing	29
	3.2.1.3 Characterization	32
	3.2.2 In-Depth Analysis of Best Placement of Carbon Source for Graphene Synthesis	36
	3.2.2.1 Sample preparation using PVD method	36

		3.2.2.2	Post-PVD annealing and characterization	37
	3.3	Growth Mechanism of PVD-Based Graphene		41
		3.3.1	Effect of Amorphous Carbon Thickness	41
		3.3.2	Effect of Annealing Time	43
		3.3.3	Effect of Annealing Temperature	45
		3.3.4	Stress Analysis Using Finite Element Modeling	46
		3.3.5	Interpretation of Experimental Results with the Aid of FEM	50
	3.4	Summary		60

4. **Statistical Approach to Identify Key Growth Parameters of the Novel Graphene Growth PVD Processes** — 65
   - 4.1 Brief History and Need of DoE — 66
   - 4.2 Importance of DoE — 67
   - 4.3 Applications of DoE — 68
   - 4.4 Illustration of DoE for the Novel PVD Graphene Synthesis — 70
     - 4.4.1 Attribute-Response Factorial Design — 70
     - 4.4.2 Full-Factorial DoE Analysis — 75
     - 4.4.3 Filtered DoE Approach — 84
   - 4.5 Summary — 86

5. **Copper–Graphene Interconnect** — 91
   - 5.1 Introduction — 91
   - 5.2 Electrical and Thermal Characteristics of Graphenated Copper — 92
     - 5.2.1 Sample Description — 92
     - 5.2.2 Experimentation and Results — 92
       - 5.2.2.1 Temperature distribution measurement — 92
       - 5.2.2.2 Electrical resistivity measurement — 93
     - 5.2.3 Atomic Level Finite Element EM Modeling — 96
   - 5.3 Compatibility of Graphenated Interconnect to Current Integrated Circuit Back-End Processes — 99

	5.4	Novel Copper–Graphene–Copper Interconnect and Its Potential Performance	101
	5.5	Mechanism of Electroless Cu Deposition on Graphenated Cu	103
	5.6	Summary	107

Index   113

Preface

The current state of modern semiconductor industry is a result of tremendous amount of research and technological advancements, spanning over five decades. During this time, the world has witnessed the evolution of modern portable devices and experienced advanced electronic technology in our lives, which is exciting, but the best is yet to come. One of the most important, basic, and tangible entities of the electronic technology is integrated circuits (ICs) whose functions depend fundamentally on two main parts: semiconductor devices and interconnects between devices to form circuits. Due to the high density of devices and aggressive downscaling, very large-scale integration (VLSI) is the norm today.

VLSI interconnects are essential since billions of transistors in an IC are interconnected to form a working circuit specific to achieve certain functions. With the advent of IC technology, the line width of VLSI interconnects is shrinking and hence their reliability is deteriorating significantly. Thus, reliability of an interconnect is an important determining factor for IC reliability. One of the major reliability concerns in IC chip level interconnection is electromigration (EM).

Copper (Cu) has remained a prime choice for an interconnection material in the semiconductor industry for years owing to its best balance of conductivity and performance. However, Cu interconnects are running out of steam as indicated by the International Technology Roadmap for Semiconductors (ITRS). Several alternatives have been suggested for Cu replacement by the ITRS, but complete replacement of Cu is not trivial due to many considerations. There have been proposals for using carbon nanotubes, 2D materials such as graphene, etc. as alternatives, but these proposals presented difficult challenges to be answered. Scalability, manufacturability, and ability to be integrated in current semiconductor process are some of the major challenges.

Owing to the excellent properties of graphene, including electrical and thermal conductivities, high current density tolerance, and high electromigration reliability on graphene interconnects, graphene is likely to be the most suitable candidate for future interconnections.

The purpose of this book is to investigate methods and techniques to fabricate "graphen(e)"ated copper (Gr-Cu) for applications of VLSI interconnects, which maintains the sanity of copper while employs the advantages of graphene. This book is a compilation of comprehensive studies of the properties of graphene and graphene synthesis methods for VLSI interconnects applications. The development of a new method to synthesize graphene and investigation of the synthesis mechanism, evaluation of electrical and electromigration performance of graphenated Cu interconnects, identification of key growth process parameters using statistical approach, and a new graphene-based electroless deposition method for copper and other metals have been covered.

This book proposes methods to fabricate graphenated interconnect systems, which have a great potential to revolutionize the VLSI-IC industry and appeal for further advancement of the semiconductor industry.

Cher-Ming Tan
Udit Narula
Vivek Sangwan
June 2021

Chapter 1

Introduction

1.1 Importance of Interconnections in VLSI

Advancements in the semiconductor industry have facilitated billions of transistors on a single chip. These billions of transistors occupy less area and enable the system to operate with higher speed and more functionality in comparison to their previous generations. These transistors must be connected to each other to form circuits with interconnects at different levels in the chip.

An aggressive increase in the number of transistors on a chip has significantly reduced the thickness and width of the interconnects. As the functionality of the chips is shifting to high-frequency and high-power applications, there is a need to give more attention to the interconnects as they can affect the speed and power consumption of the chip. With the increase in the number of interconnections due to their need to connect an ever-increasing number of transistors in a chip, and the space available for interconnections is reducing, the cross-sectional area of interconnects needs to be reduced, but this will increase the resistance of the interconnects and the current density in the interconnects that reduces their speed and electromigration performances.

According to the International Technology Roadmap for Semiconductors (ITRS), the lifetime of very large-scale integration

---

Graphene and VLSI Interconnects
Cher-Ming Tan, Udit Narula, and Vivek Sangwan
Copyright © 2022 Jenny Stanford Publishing Pte. Ltd.
ISBN 978-981-4877-82-4 (Hardcover), 978-1-003-22488-4 (eBook)
www.jennystanford.com

(VLSI) interconnects will decrease by half with the new generation [1]. One of the most prominent limiting factors for this is electromigration (EM). Hence, a detailed study is necessary to ensure better interconnect systems with the aim of enhancing the overall reliability of the circuit, and this calls for a detailed basic understanding of the interconnects with regard to its EM performances.

1.2   Copper Is Running Out of Steam

Initial studies on the EM failure mechanism were reported for copper [2]; however, more effort was put into researching the EM phenomenon of aluminum [3, 4] subsequently since the interconnect material for VLSI was then Al.

However, the aggressive scaling of VLSI has challenged the reliability of Al interconnects, and the modification of Al interconnects such as aluminum–silicon (Al–Si), aluminum–copper (Al–Cu), etc. [5–7] was employed to resolve contact spiking and improve Al interconnect EM performance. With the increasing demand for miniaturization, Al-based interconnections were proved to be inefficient in the 1990s due to their higher electrical resistivity and thus insufficient circuit speed, and the choice of alternative material for interconnect was made where copper was selected even though it has the second lowest electrical resistivity next to silver. Silver is not chosen because it is prone to electrochemical migration [8]. Today, the semiconductor industry uses copper predominantly.

With continuous scaling, even copper interconnects are now running out of steam as mentioned in the ITRS [9]. The roadmap laid down critical challenges for interconnects [9] for the technology node beyond 16 nm. It pointed out that improvement in the following key aspects can bring about a better interconnect in ICs.

- Choice of the material of interconnections
- Reliable performance of interconnects
- Manufacturability of the new interconnect materials
- Cost and yield of the interconnect materials

Among many reliability issues in IC interconnects, EM is one of the important reliability concerns [10, 11]. EM has become the major concern for the reliability of IC interconnects due to the constantly decreasing cross-sectional area of interconnects, which results in increased operational current density that is extremely high for shrinking interconnects [12], increasing the severity of the EM-induced mass transport through diffusion [13, 14].

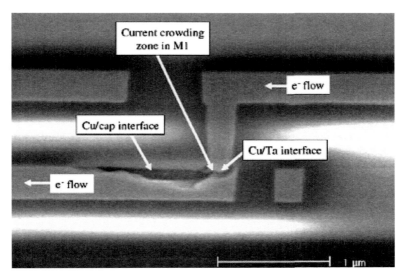

Figure 1.1 SEM image of a failed sample. Reprinted with the permission from Ref. [17], Copyright 2005, IEEE.

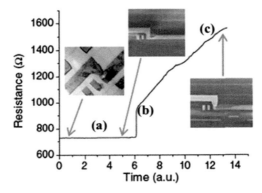

Figure 1.2 Typical package-level EM resistance trace. Reprinted with the permission from Ref. [18], Copyright 2008, AIP Publishing.

EM creates voids in interconnections that lead to open circuit or increased line resistance, and consequent circuit failure [15]. There might also be hillocks formation that results in short circuit between adjacent interconnects [16]. Figure 1.1 depicts the SEM image of a failed sample as obtained by Tan et al. [17]. A void can be clearly seen there.

Figure 1.2 depicts the EM degradation process, which explains its effect on the resistance of an interconnect.

1.3 Graphene as a Solution

The ITRS suggested some alternatives to copper for interconnects [9], and the information is extracted and summarized in Table 1.1, with their respective associated primary concerns.

Table 1.1 Copper replacement suggested by ITRS

Option	Potential Advantages	Primary Concerns
Other metals (silver, silicide, stacks)	Potential lower resistance in fine geometries	Grain boundary scattering, integration issues, reliability
Nanowires	Ballistic conduction in narrow lines	Quantum contact resistance, controlled placement, low density, substrate interactions
Carbon nanotubes	Ballistic conduction in narrow lines, electromigration resistance	Quantum contact resistance, controlled placement, low density, chirality control, substrate interactions, parametric spread
Graphene nanoribbons	Ballistic conduction in narrow films, planar growth, electromigration resistance	Quantum contact resistance, control of edges, deposition, etch stopping and stacking substrate interactions
Optical (interchip)	High bandwidth, low power and latency, noise immunity	Connection and alignment between die and package, optical/electrical conversion efficiencies

Option	Potential Advantages	Primary Concerns
Optical (intrachip)	Latency and power reduction for long lines, high bandwidth with wavelength-division multiplex (WDM)	Benefits only for long lines, need compact components, integration issues, need WDM, energy cost
Wireless	Available with current technology, parallel transport medium, high fan-out capability	Very limited bandwidth, intra-die communication difficult, large area and power overhead
Superconductors	Zero-resistance interconnect, high Q passives	Cryogenic cooling, frequency-dependent resistance, defects, low critical current density, inductive noise, and crosstalk

In Table 1.1, the suggested optical and wireless options require different chip designs, while superconductor option requires a cryogenic environment. Therefore, these options will be given lower priority as their implementation will require a lot more work. On the other hand, the current chip design methodology may be retained if silver, nanowires, carbon nanotubes, and/or graphene nanoribbons (GNRs) are used, but silver has a higher electrochemical migration problem as mentioned earlier and nanowires have placement and contact resistance issues just like carbon nanotubes.

The high contact resistance with copper, which is in the range of a few mega-ohms for single-walled nanotubes [19, 20] and up to several mega-ohms for multi-walled nanotubes [21], has rendered them inefficient for copper replacement and stacking with copper for multi-level interconnection. GNR alone is also difficult to stack in multi-level metallization although the contact resistance of graphene with copper is in the range of a few kilo-ohms [22]. Hence, it is clear from the preceding discussion that complete replacement of copper requires more research. Owing to the excellent material properties of graphene, it could be a good choice to replace copper interconnections.

Graphene has been a hot topic in material science because of its single atomic thickness and yet excellent electrical and thermal conductivities alongside other superior mechanical and optical properties, which will be discussed in the next section. The most intriguing property of graphene, which forms the most important motivation for this book, is the high current density tolerance for pristine graphene, which is stated to be 10^8 A/cm^2 [23, 24]. This value is almost two-three orders higher than copper. There have been reliability studies on graphene interconnects, which give a new dimension to research on interconnects [25, 26], as depicted in Figs. 1.3–1.5.

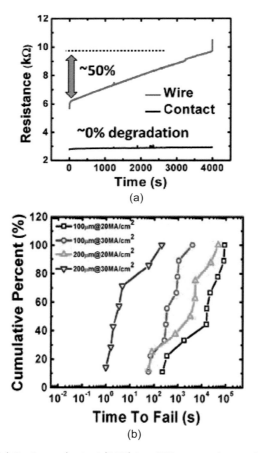

Figure 1.3 (a) Device under test (DUT) is a 100 μm graphene wire in contact with Ti/Au electrodes, tested at 30 MA/cm^2 at 523 K. (b) Graphene interconnect lifetime distribution at 523 K for different wire lengths and stress current densities. Each of the lines in the figure shows the lifetime of approximately 10 wires, each tested individually. Reprinted with the permission from Ref. [25], Copyright 2012, IEEE.

Graphene as a Solution | 7

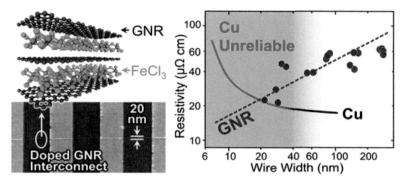

Figure 1.4 Intercalation doped multilayer-graphene-nanoribbons. Reprinted with permission from Ref. [26], Copyright 2017, American Chemical Society.

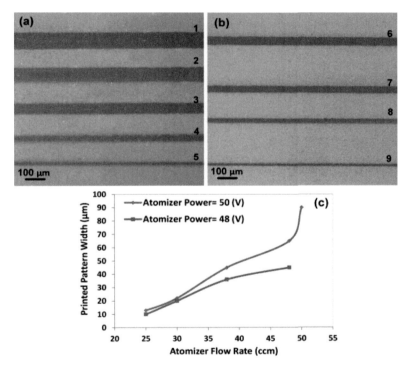

Figure 1.5 Microscale aerosol-jet printing of graphene interconnects. Reprinted from Ref. [27], Copyright 2015, with permission from Elsevier.

From Fig. 1.3a, it is clear that the graphene wire's resistance degrades linearly with time and ultimately the breakdown of the wire is observed under constant current stress of 30 MA/cm² while there is no significant change in the contact resistance between graphene and Ti/Au used as contact electrodes. This kind of superior stable contact resistance is believed to be similar for a Cu–graphene contact, and it can be a good condition for a graphene–Cu interconnect. Figure 1.3b shows that the EM performance of graphene is excellent.

A comparison of the data in Fig. 1.3 with copper as extracted from the EM tests on copper shown in Fig. 1.6 is summarized in Table 1.2.

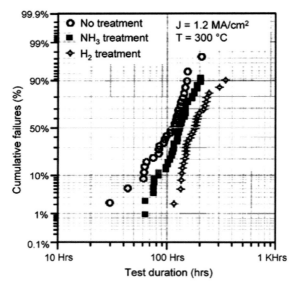

Figure 1.6 EM performance comparison for copper. Reprinted from Ref. [28], Copyright 2015, with permission from Elsevier.

It is clear from Table 1.2 that the t_{50} for graphene for different current densities is much better than copper. Hence, graphene has the potential to be used as interconnects in the future.

However, fabrication and handling of graphene interconnects are challenging and do not fit into the conventional back-end-of-line (BEOL) operations [29]. There are many reasons for this, and some of them are the introduction of defects in graphene during the transfer process and photolithographic treatments, and inevitable nano-holes creation during fabrication. Hence, it is more logical to use graphene with copper as a combined entity.

Table 1.2 Comparison of plots in Fig. 1.3 and Fig. 1.6

Figure [Reference]	DUT Length (μm)	Stress Conditions Current Density (MA/cm^2)	Temperature (K)	Time for 50% Degradation (Hours)
1.3 [25]	100	20	523	10
1.6 [28]	800	1.2	523	100
Computed t_{50} for copper using Black's equation [3] assuming same activation energy E_a (in reality; E_a for graphene should be higher than Cu)		20	523	6 ($n=1$) 0.36 ($n=2$)

Note: n is the current density exponent.

Therefore, the motivation for the book comes from graphene as a candidate to improve and push the limits of copper further as IC technology is continuing to shrink. Information and important properties of graphene are thus articulated in the following section.

1.4 Properties and Applications of Graphene

Graphene is a nanomaterial that has gained immense popularity among academicians, researchers, and industries since its discovery in 2004 [29]. By virtue of its fantastic electrical, thermal, and mechanical properties, graphene has achieved pre-eminence in the field of nanotechnology. Some of the worth-noting properties of graphene are as follows [30]:

- Single atomic graphite layer thickness [29], which is classified as a quasi-metal because of its semiconducting metal properties, even though technically it is a nonmetal. It exhibits sp^2 hybridization where a covalent bond is present in each carbon atom with other three carbon atoms in a hexagonal array fashion. This makes one free electron responsible for electronic conduction.
- High current density capacity, i.e., 10^8 A/cm^2 [31], which is 100 times larger than copper.

- Extremely high mobility of about 15,000 cm^2/Vs [32, 33], which is independent of room temperature. As graphene is a zero-bandgap material, it is difficult to turn off the device completely. This makes it near to impossible to use it in logic circuits such as field-effect transistors (FETs). This limitation can be overcome by widening the bandgap by the use of quantum confinement, which results in GNRs.
- High thermal conductivity in the order of 5×10^3 W/mK [34], which is 10× as compared to copper. This enables it useful for high-current applications where high temperature is usually generated.
- Almost transparent material, possess high optical transmittance of about 97% within the visible light wavelength range of 400 to 700 nm [35].
- High mechanical strength with breaking strength of 42 N/m [36], which is 200 times more than steel. Its second- and third-order elastic stiffness values are 340 N/m and −690 N/m, respectively. It is also known as the world's strongest material.

Application of graphene in humanity is intriguing, and it has emerged as an encouraging contender for numerous applications. In biomedicine applications, it can be used as an efficient nanocarrier for drug delivery [37], gene delivery [38], cancer therapy [39], antibacterial materials [40], and biological imaging [41]. Graphene-based FETs can be employed for chemical and biosensors due to graphene's highly sensitive electrical property [42].

Graphene-based materials such as photocatalytic or sorbent materials can be used in environmental decontamination applications [43]. It is used in the manufacturing of desalination membranes and water treatment. It can be utilized as an electrode material for contaminant monitoring or removal. As an absorbent, it can be used as either metal ion adsorption or organic compound adsorption.

It is also playing a major role in developing a new kind of energy storage devices because of its excellent electrical and mechanical properties in addition to its large surface area [44]. It, therefore, enables the capacity enhancement of supercapacitors that are also known as electrochemical capacitors and ultracapacitors, by using it as a supercapacitor electrode.

Graphene-based metal nanostructures are popular and are used in the development of fuel cells [45], batteries [46], photovoltaic devices [47], photocatalysts [48], and electrochemical sensors [49]. Facilitation in the enhanced performance of the devices attributed due to the large surface area and high conductivity of graphene, which in return leads to effective ion and electron transport.

Graphene is also involved in the field of electronic devices for radio frequency (RF) applications. The main use of graphene can be observed in the development of low noise amplifiers (LNAs) [50], nonlinear electronics [51], RF electromechanical resonators [52], and switches [53]. For example, in HF LNA, graphene high mobility charge carriers provide a low resistance path between the source to drain that helps to achieve high current densities and high operational efficiency. The short channel effect can also be reduced because of its one-atom thickness and enables it for use in ultra-high frequency operations [54].

1.5 Summary

This chapter provides an overview of the importance of interconnections in VLSI technology. As the technology progresses, the size of interconnects is shrinking, which adversely affects the reliability and functionality of the system. In order to overcome these limitations, different kinds of materials are proposed in order to meet the future needs of the interconnects, and graphene comes out to be one of the most important contenders due to its versatile advantageous material properties, and various applications of graphene have been implemented as discussed in this chapter.

References

1. "International Technology Roadmap for Semiconductors," 2014. [Online]. Available: available: http://public.itrs.net.
2. A. R. Grone, "Current-induced marker motion in copper," *J. Phys. Chem. Solids*, **20**, 1–2, pp. 88–93, 1961.
3. J. R. Black, "Electromigration: A brief survey and some recent results," *IEEE Trans. Electron Devices*, **16**, 4, pp. 338–347, 1969.

4. S. M. Spitzer and S. Schwartz, "The effects of dielectric overcoating on electromigration in aluminum interconnections," *IEEE Trans. Electron Devices*, **16**, 4, pp. 348–350, 1969.
5. I. A. Blech, "Copper electromigration in aluminum," *J. Appl. Phys.*, **48**, 2, pp. 473–477, 1977.
6. S. Vaidya, D. B. Fraser, and W. S. Lindenberger, "Electromigration in fine-line sputter-gun Al," *J. Appl. Phys.*, **51**, 8, pp. 4475–4482, 1980.
7. J. K. Howard, J. F. White, and P. S. Ho, "Intermetallic compounds of Al and transitions metals: Effect of electromigration in 1-2-µm-wide lines," *J. Appl. Phys.*, **49**, 7, pp. 4083–4093, 1978.
8. A. Christou, *Electromigration & Electronic Device degradation*, Wiley, 1993.
9. ITRS, "Interconnect Summary," 2013.
10. C. M. Tan and A. Roy, "Electromigration in ULSI interconnects," *Mater. Sci. Eng. R Reports*, **58**, 1–2, pp. 1–75, 2007.
11. L. D. Chen et al., "Study of a new electromigration failure mechanism by novel test structure," in *IEEE International Reliability Physics Symposium Proceedings*, 2015, pp. 2D5.1–2D5.5.
12. B. Geden, "Understand and avoid electromigration (EM) & IR-drop in custom IP blocks," in *Synopsys White Paper*, 2011, pp. 1–6.
13. S. Yokogawa and H. Tsuchiya, "Scaling impacts on electromigration in narrow single-damascene Cu interconnects," *Japanese J. Appl. Physics, Part 1 Regul. Pap. Short Notes Rev. Pap.*, **44**, 4A, pp. 1717–1721, 2005.
14. F. Chen et al., "Diagnostic electromigration reliability evaluation with a local sensing structure," in *IEEE International Reliability Physics Symposium Proceedings*, 2015, pp. 2D4.1–2D4.7.
15. C. M. Tan, *Electromigration in ULSI Interconnections*, World Scientific, 2010.
16. A. Gladkikh, Y. Lereah, E. Glickman, M. Karpovski, A. Palevski, and J. Schubert, "Hillock formation during electromigration in Cu and Al thin films: Three-dimensional grain growth," *Appl. Phys. Lett.*, **66**, 10, pp. 1214–1215, 1995.
17. C. M. Tan, A. Roy, A. V. Vairagar, A. Krishnamoorthy, and S. G. Mhaisalkar, "Current crowding effect on copper dual damascene via bottom failure for ULSI applications," *IEEE Trans. Device Mater. Reliab.*, **5**, 2, pp. 198–205, 2005.
18. L. Doyen, E. Petitprez, P. Waltz, X. Federspiel, L. Arnaud, and Y. Wouters, "Extensive analysis of resistance evolution due to electromigration induced degradation," *J. Appl. Phys.*, **104**, 12, 2008.

19. L. An and C. R. Friedrich, "Measurement of contact resistance of multiwall carbon nanotubes by electrical contact using a focused ion beam," *Nucl. Instrum. Methods Phys. Res., B*, **272**. pp. 169–172, 2012.
20. K. Kong, S. Han, and J. Ihm, "Development of an energy barrier at the metal-chain–metallic-carbon-nanotube nanocontact," *Phys. Rev. B Condens. Matter Mater. Phys.*, **60**, 8, pp. 6074–6079, 1999.
21. M. Park *et al.*, "Effects of a carbon nanotube layer on electrical contact resistance between copper substrates," *Nanotechnology*, **17**, 9, pp. 2294–2303, 2006.
22. S. K. Hong, S. M. Song, O. Sul, and B. J. Cho, "Reduction of metal-graphene contact resistance by direct growth of graphene over metal," *Carbon Lett.*, **14**, 3, pp. 171–174, 2013.
23. R. Murali, Y. Yang, K. Brenner, T. Beck, and J. D. Meindl, "Breakdown current density of graphene nanoribbons," *Appl. Phys. Lett.*, **94**, 24, 2009.
24. T. Yu, E. Lee, B. Briggs, B. Nagabhirava, and B. Yu, "Bilayer graphene/copper hybrid on-chip interconnect: A reliability study," *IEEE Trans. Nanotechnol.*, **10**, 4, pp. 710–714, 2011.
25. X. Chen, D. H. Seo, S. Seo, H. Chung, and H.-S. P. Wong, "Graphene interconnect lifetime: A reliability analysis," *IEEE Electron Device Lett.*, **33**, 11, pp. 1604–1606, 2012.
26. J. Jiang *et al.*, "Intercalation doped multilayer-graphene-nanoribbons for next-generation interconnects," *Nano Lett.*, **17**, 3, pp. 1482–1488, 2017.
27. E. Jabari and E. Toyserkani, "Micro-scale aerosol-jet printing of graphene interconnects," *Carbon N. Y.*, **91**, pp. 321–329, 2015.
28. A. V. Vairagar, S. G. Mhaisalkar, and A. Krishnamoorthy, "Effect of surface treatment on electromigration in sub-micron Cu damascene interconnects," *Thin Solid Films*, **462–463**, pp. 325–329, 2004.
29. D. Johnson, "Carbon nanomaterials could push copper aside in chip interconnects," *IEEE Spectr.*, 2017.
30. K. S. Novoselov *et al.*, "Electric field effect in atomically thin carbon films," *Science*, **306**, 5696, pp. 666–669, 2004.
31. U. Narula, C. M. Tan, and C. S. Lai, "Growth mechanism for low temperature PVD graphene synthesis on copper using amorphous carbon," *Sci. Rep.*, **7**, pp. 1–13, 2017.
32. J. Moser, A. Barreiro, and A. Bachtold, "Current-induced cleaning of graphene," *Appl. Phys. Lett.*, **91**, 16, pp. 1–4, 2007.

33. A. K. Geim and K. S. Novoselov, "The rise of graphene," *Nat. Mater.*, **6**, 3, pp. 183–191, 2007.
34. K. I. Bolotin *et al.*, "Ultrahigh electron mobility in suspended graphene," *Solid State Commun.*, **146**, 9–10, pp. 351–355, 2008.
35. A. A. Balandin *et al.*, "Superior thermal conductivity of single-layer graphene," *Nano Lett.*, **8**, 3, pp. 902–907, 2008.
36. R. R. Nair *et al.*, "Fine structure constant defines visual transparency of graphene," *Science*, **320**, 5881, p. 1308, 2008.
37. C. Lee, X. Wei, J. W. Kysar, and J. Hone, "Measurement of the elastic properties and intrinsic strength of monolayer graphene," *Science*, **321**, 5887, pp. 385–388, 2008.
38. D. Depan, J. Shah, and R. D. K. Misra, "Controlled release of drug from folate-decorated and graphene mediated drug delivery system: Synthesis, loading efficiency, and drug release response," *Mater. Sci. Eng. C*, **31**, 7, pp. 1305–1312, 2011.
39. H. Bao *et al.*, "Chitosan-functionalized graphene oxide as a nanocarrier for drug and gene delivery," *Small*, **7**, 11, pp. 1569–1578, 2011.
40. K. Yang, L. Feng, and Z. Liu, "Stimuli responsive drug delivery systems based on nano-graphene for cancer therapy," *Adv. Drug Deliv. Rev.*, **105**, pp. 228–241, 2016.
41. W. Hu *et al.*, "Graphene-based antibacterial paper," *ACS Nano*, **4**, 7, pp. 4317–4323, 2010.
42. J. L. Li, B. Tang, B. Yuan, L. Sun, and X. G. Wang, "A review of optical imaging and therapy using nanosized graphene and graphene oxide," *Biomaterials*, **34**, 37, pp. 9519–9534, 2013.
43. C. S. Park, H. Yoon, and O. S. Kwon, "Graphene-based nanoelectronic biosensors," *J. Ind. Eng. Chem.*, **38**, pp. 13–22, 2016.
44. H. Wang *et al.*, "Graphene-based materials: Fabrication, characterization and application for the decontamination of wastewater and waste gas and hydrogen storage/generation," *Adv. Colloid Interface Sci.*, **195–196**, pp. 19–40, 2013.
45. W. Lv, Z. Li, Y. Deng, Q. H. Yang, and F. Kang, "Graphene-based materials for electrochemical energy storage devices: Opportunities and challenges," *Energy Storage Mater.*, **2**, pp. 107–138, 2016.
46. M. V. Kannan and G. Gnana Kumar, "Current status, key challenges and its solutions in the design and development of graphene based ORR catalysts for the microbial fuel cell applications," *Biosens. Bioelectron.*, **77**, pp. 1208–1220, 2016.

47. M. Pumera, "Graphene-based nanomaterials for energy storage," *Energy Environ. Sci.*, **4**, 3, pp. 668–674, 2011.
48. Z. Yin *et al.*, "Organic photovoltaic devices using highly flexible reduced graphene oxide films as transparent electrodes," *ACS Nano*, **4**, 9, pp. 5263–5268, 2010.
49. Q. Xiang, J. Yu, and M. Jaroniec, "Graphene-based semiconductor photocatalysts," *Chem. Soc. Rev.*, **41**, 2, pp. 782–796, 2012.
50. X. Kang, J. Wang, H. Wu, J. Liu, I. A. Aksay, and Y. Lin, "A graphene-based electrochemical sensor for sensitive detection of paracetamol," *Talanta*, **81**, 3, pp. 754–759, 2010.
51. S. Das and J. Appenzeller, "An all-graphene radio frequency low noise amplifier," in *2011 IEEE Radio Frequency Integrated Circuits Symposium*, 2011, pp. 1–4.
52. S. Yamashita, "A tutorial on nonlinear photonic applications of carbon nanotube and graphene," *J. Light. Technol.*, **30**, 4, pp. 427–447, 2012.
53. J. S. Bunch *et al.*, "Electromechanical resonators from graphene sheets," *Science*, **315**, 5811, pp. 490–494, 2007.
54. T. Palacios, A. Hsu, and H. Wang, "Applications of graphene devices in RF communications," *IEEE Commun. Mag.*, **48**, 6, pp. 122–128, 2010.

Chapter 2

Graphene Synthesis Methods

2.1 Overview of Existing Methods

Graphene synthesis methods have witnessed many pathways in order to produce high-quality graphene with a high degree of coverage area. Beginning from mechanical exfoliation [1], graphene synthesis methods such as graphite sonication [2], epitaxial growth [3], and graphene oxide reduction [4] have been evolved in order to produce low-cost, uniform, large area, relatively low defect graphene films.

2.1.1 Mechanical Exfoliation

This method is also known as the "scotch-tape method," and this is the method through which graphene was discovered in 2004 by Novoselov et al. in the Manchester University [1]. In this method, graphene (two-dimensional crystal) is produced from graphite (three-dimensional crystal) over the substrate by peeling the bulk graphite layer by layer, where van der Waals force helps to stick this graphene layer over the substrate. It can be isolated from the substrate by the etching process in order to produce self-supporting graphene.

Two kinds of mechanical methods are available to exfoliate graphene from graphite: normal force method and lateral force

Graphene and VLSI Interconnects
Cher-Ming Tan, Udit Narula, and Vivek Sangwan
Copyright © 2022 Jenny Stanford Publishing Pte. Ltd.
ISBN 978-981-4877-82-4 (Hardcover), 978-1-003-22488-4 (eBook)
www.jennystanford.com

method [5]. When an external force is applied in the normal direction to overcome the van der Waals attraction at the graphite to produce graphene, it is known as the normal force method. On the other hand, in the lateral force method, a lateral force is applied at the graphite to generate graphene. The feasibility of this method depends on the intrinsic material property of graphene, i.e., self-lubricating.

2.1.2   Graphite Sonication

In this method, acoustic or ultrasonic waves are used to exfoliate graphite powders that are dissolved in the solvents with the aim of producing graphene [6]. The waves propagate through graphite and develop an alternating high- and low-pressure cycles, where each of these cycles plays its own major role in the production of graphene. In the low-pressure cycle, small vacuum bubbles are created, and in the consecutive high-pressure cycle, these small bubbles absorb the energy and burst. This results in the cavitation phenomenon, which produces high-velocity jets and shock waves simultaneously, creating both normal and lateral forces over the graphite, and helps to exfoliate the graphite, resulting in graphene layers.

Different kinds of solvents can be used in this process, such as water, N-methyl-pyrrolidone (NMP), or N-dimethyl-formamide (DMF) [7]. NMP has some good advantages over its counterpart in that it develops defect-free graphene because its surface energy is well-matched with graphene, helping the exfoliation process to occur freely. However, NMP is expensive to use and special care is required while handling because it possesses high boiling point that makes the deposition of monolayer graphene difficult. On the other hand, one of the most useful and popular solvents, i.e., water has some drawbacks such as its very high surface energy that makes it difficult to exfoliate graphite to produce graphene.

2.1.3   Graphene Oxide Reduction

Graphene obtained from this process is known as reduced graphene oxide (rGO). However, rGO does not reach its full potential as compared to the graphene obtained from various methods. Nevertheless, with its valuable properties, it can be used in multiple applications because of its unique optical, electrical, and thermal properties [8].

The precursor for rGO is graphite oxide, and it is synthesized by the oxidation of graphite through various chemical ways. A few of the popular chemical methods are Brodie's oxidation method [9], Staudenmaier method [10], Hofmann method [11], and Hummers method [12]. During the oxidation process, oxidizing agents react with graphite and produce graphite oxide. Chemically obtained graphite oxide and graphene oxide are similar; however, the major difference between them is in their respective structures. In graphene oxide, the interplanar spacing between the two consecutive atomic layers is larger and the disrupted sp^2 bonding network is found as compared to graphite oxide, due to the oxidation process.

The oxidizing process is followed by the dispersion of the obtained oxidized compound in a base solution, and graphene oxide is then obtained. Reduction of graphite oxide to rGO can be obtained in multiple ways such as chemical reduction [13], thermal reduction [14], microwave- and photo-reduction [15, 16], photocatalyst reduction [17], and solvothermal/hydrothermal reduction [18].

2.1.4 Epitaxial Growth

Graphene can be produced epitaxially over the transition metal surfaces through the vapor deposition process, which is based on the transfer of material in the vapor phase at either atomic or molecular level. It can be performed via different methods such as chemical vapor decomposition (CVD) or physical vapor decomposition (PVD) method. However, parameters such as thin film material, the thickness of the deposited layer, and substrate type define the deposition method.

One of the most current popular methods is the CVD method [19–22], which is known to be the cheapest for mass production. In this method, chemical reaction needs to happen either amid two reactive gases or between the substrate and reactant. A dedicated ultra-high vacuum (UHV) chamber is required with a pressure of 10–11 mbar and gas sputter (around 1 keV). The sample is kept at a very high temperature, generally more than 700 K followed by the deposition of coating material molecules over the substrate surface. A gas phase is used in this process for surface coating, and it is appropriate for the coating of large and complex-shaped surfaces.

The PVD method uses physical processes for the thin film decomposition over the substrate surface, which can be grown by thermal evaporation, anodic arc evaporation, or sputtering physical processes. This process is performed under the low vacuum and low temperature to deposit the condensed gas-like material over the substrate and the process is known as an atomistic deposition process. The advantage of the vacuum environment is that the metals will not react with the atmospheric oxygen, which results in high quality of the deposited layer.

2.2 Drawbacks of Current Methods with Respect to Applications in VLSI Interconnect Fabrication

Graphene obtained from the mechanical exfoliation method is of good quality; the preparation process is exceptionally simple, and the complete fabrication process finishes in a few minutes. However, the drawback of this method is that the graphene deposition growth cannot be controlled in real time, which renders mass production and selective placement of graphene impossible, and it is, therefore, not suitable in the VLSI interconnect fabrication.

The sonication method produces graphene of high purity with high yield. The major challenges are to keep the medium surface energy as close to that of graphene, i.e., 68 mJ/m^2 [23], and maintain good physical interaction strength between the medium and graphene for the proper deposition. This method requires the use of surfactants, acids, chemical reagents, and organic solvents in the manufacturing process. Although all of them help to enhance the quality and yield of graphene, they produce many challenges such as high toxicity, high cost, difficulty to remove the surfactants residuals, and negative impact of these reagents on the environment. All these are the major challenges for the VLSI fabrication.

Although good quality of graphene can also be obtained through the rGO method, it possesses some drawbacks while using it for the VLSI interconnects. If rGO is used over copper substrate, the defects density of rGO is found to increase between 0 V (no bias) and 0.5 V, remain constant from 0.5 V to 1.5 V, and later decrease with the increase in the applied bias for the test sample from 1.5 V to 3 V [24].

It is verified by the I_D/I_G ratio as depicted in Fig. 2.1, and this variation can lead to the change in the mobility, electrical conductivity, and resistivity of the samples, which will affect the device performances.

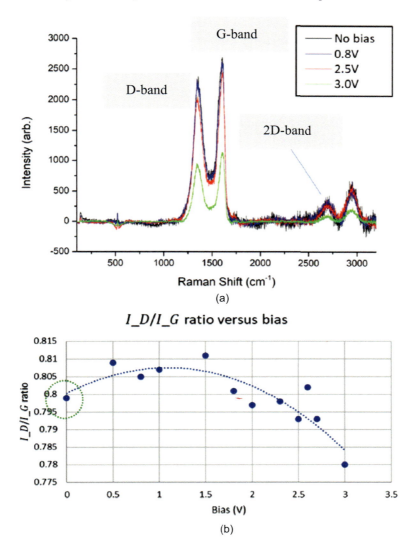

Figure 2.1 (a) Raman spectra of graphene oxide under different voltages. (b) The I_D/I_G ratio of GO sample after electrical stimulus. The no-bias point is marked by the dotted-circle line. Republished with permission of Inderscience Enterprises Limited (UK), from Ref. [24]; permission conveyed through Copyright Clearance Center, Inc.

Since there is a plethora of applications for copper material, growth of graphene on copper can overcome some known limitations of Cu material such as its ease of corrosion, ease of oxidation, limited electromigration performance as interconnect, and the electrical and thermal conductivity limits of Cu [25].

Out of all the existing graphene growth methods, the CVD method seems to be most suitable for the said application due to the low cost for mass production and better quality in terms of size and properties [26].

However, on careful studying and experimenting with the CVD method, it has the following drawbacks with respect to application in VLSI interconnects:

- The process temperature is still too high (in the range of 800–1025°C) when it comes to the application of graphene in the semiconductor industry.
- The process results in transferable graphene, which is prone to damage on transfer.
- The synthesis is mostly carried out on copper foils and not on ultra-thin copper film, which is a requirement for interconnections.
- A lot of gaseous carbon source is wasted during the beginning and end of the process.

Hence, it becomes necessary to establish a non-transferable graphene synthesis method, which has the potential to decrease the synthesis temperature to a suitable value. Non-transferable graphene synthesis is important especially for VLSI interconnects, which are fabricated in the back-end-of-line (BEOL) process, and transfer of graphene is infeasible [25]. Moreover, the transfer of graphene introduces many structural inhomogeneities that cause degradation of properties of graphene such as mobility, electrical conductivity, and thermal conductivity [27–29].

2.3 Summary

This chapter highlights the fascinating and magnificent properties of graphene, which has huge prospects for many applications. VLSI interconnects are the main focus of this work. In this chapter, the graphene synthesis methods have been studied in detail with respect to the suitability of the method to the VLSI interconnect fabrication application. The pros and cons of the methods have been explored, which leads to the need for a new graphene synthesis method that can be applied to standard interconnect fabrication process and will be discussed in the next chapter.

References

1. K. S. Novoselov et al., "Electric field effect in atomically thin carbon films," *Science*, **306**, 5696, pp. 666–669, 2004.
2. K. H. Choi, A. Ali, and J. Jo, "Randomly oriented graphene flakes film fabrication from graphite dispersed in N-methyl-pyrrolidone by using electrohydrodynamic atomization technique," *J. Mater. Sci. Mater. Electron.*, **24**, 12, pp. 4893–4900, 2013.
3. P. Macháč, T. Fidler, S. Cichoň, and V. Jurka, "Synthesis of graphene on Co/SiC structure," *J. Mater. Sci. Mater. Electron.*, **24**, 10, pp. 3793–3799, 2013.
4. M. I. Ali Umar, C. C. Yap, R. Awang, M. H. Hj Jumali, M. Mat Salleh, and M. Yahaya, "Characterization of multilayer graphene prepared from short-time processed graphite oxide flake," *J. Mater. Sci. Mater. Electron.*, **24**, 4, pp. 1282–1286, 2013.
5. M. Yi and Z. Shen, "A review on mechanical exfoliation for the scalable production of graphene," *J. Mater. Chem. A*, **3**, 22, pp. 11700–11715, 2015.
6. K. Muthoosamy and S. Manickam, "State of the art and recent advances in the ultrasound-assisted synthesis, exfoliation and functionalization of graphene derivatives," *Ultrason. Sonochem.*, **39**, pp. 478–493, 2017.
7. S. Vadukumpully, J. Paul, and S. Valiyaveettil, "Cationic surfactant mediated exfoliation of graphite into graphene flakes," *Carbon N. Y.*, **47**, 14, pp. 3288–3294, 2009.
8. R. Bhargava and S. Khan, "Effect of reduced graphene oxide (rGO) on structural, optical, and dielectric properties of $Mg(OH)_2$/rGO nanocomposites," *Adv. Powder Technol.*, **28**, 11, pp. 2812–2819, 2017.
9. B. C. Brodie, "On the atomic weight of graphite," *R. Soc. London*, **149**, pp. 249–259, 1858.
10. L. Staudenmaier, "Method for the preparation of the graphite acid," *Eur. J. Inorg. Chem.*, **31**, 2, pp. 1481–1487, 1898.

11. U. Hofmann and A. Frenzel, "Die Reduktion von Graphitoxyd mit Schwefelwasserstoff," *Kolloid-Zeitschrift*, **68**, 2, pp. 149–151, 1934.
12. W. S. Hummers and R. E. Offeman, "Preparation of graphitic oxide," *J. Am. Chem. Soc.*, **80**, 6, p. 1339, 1958.
13. G. Brauer, *Handbook of Preparative Inorganic Chemistry*, vol. 1. Academic Press, 1963.
14. X. Gao, J. Jang, and S. Nagase, "Hydrazine and thermal reduction of graphene oxide: Reaction mechanisms, product structures, and reaction design," *J. Phys. Chem. C*, **114**, 2, pp. 832–842, 2010.
15. H. M. A. Hassan et al., "Microwave synthesis of graphene sheets supporting metal nanocrystals in aqueous and organic media," *J. Mater. Chem.*, **19**, 23, pp. 3832–3837, 2009.
16. Y. Matsumoto et al., "Simple photoreduction of graphene oxide nanosheet under mild conditions," *ACS Appl. Mater. Interfaces*, **2**, 12, pp. 3461–3466, 2010.
17. Y. H. Ng, A. Iwase, A. Kudo, and R. Amal, "Reducing graphene oxide on a visible-light $BiVO_4$ photocatalyst for an enhanced photoelectrochemical water splitting," *J. Phys. Chem. Lett.*, **1**, 17, pp. 2607–2612, 2010.
18. T. Qi, C. Huang, S. Yan, X. J. Li, and S. Y. Pan, "Synthesis, characterization and adsorption properties of magnetite/reduced graphene oxide nanocomposites," *Talanta*, **144**, pp. 1116–1124, 2015.
19. S. Thiele et al., "Engineering polycrystalline Ni films to improve thickness uniformity of the chemical-vapor-deposition-grown graphene films," *Nanotechnology*, **21**, 1, 2010.
20. A. Reina et al., "Large area, few-layer graphene films on arbitrary substrates by chemical vapor deposition," *Nano Lett.*, **9**, 1, pp. 30–35, 2009.
21. A. Reina et al., "Growth of large-area single- and Bi-layer graphene by controlled carbon precipitation on polycrystalline Ni surfaces," *Nano Res.*, **2**, 6, pp. 509–516, 2009.
22. X. Li et al., "Large-area synthesis of high-quality and uniform graphene films on copper foils," *Science*, **324**, 5932, pp. 1312–1314, 2009.
23. J. N. Coleman, "Liquid exfoliation of defect-free graphene," *Acc. Chem. Res.*, 46, 1, pp. 14–22, 2013.
24. Z. Zeng, P. Singh, S. L. Xiaodai, C. M. Tan, and C. H. Sow, "In-situ characterization of the defect density in reduced graphene oxide under different electrical stress using fluorescence microscopy," *Int. J. Nanotechnol.*, **17**, 1, pp. 57–70, 2020.

25. U. Narula, C. M. Tan, and C. S. Lai, "Growth mechanism for low temperature PVD graphene synthesis on copper using amorphous carbon," *Sci. Rep.*, **7**, pp. 1–13, 2017.
26. K. S. Novoselov, V. I. Fal'Ko, L. Colombo, P. R. Gellert, M. G. Schwab, and K. Kim, "A roadmap for graphene," *Nature*, **490**, 7419, pp. 192–200, 2012.
27. B. J. Park et al., "Realization of large-area wrinkle-free monolayer graphene films transferred to functional substrates," *Sci. Rep.*, **5**, p. 9610, 2015.
28. A. Pirkle et al., "The effect of chemical residues on the physical and electrical properties of chemical vapor deposited graphene transferred to SiO_2," *Appl. Phys. Lett.*, **99**, 12, pp. 2009–2012, 2011.
29. V. E. Calado, G. F. Schneider, A. M. M. G. Theulings, C. Dekker, and L. M. K. Vandersypen, "Formation and control of wrinkles in graphene by the wedging transfer method," *Appl. Phys. Lett.*, **101**, 10, pp. 99–102, 2012.

Chapter 3

Novel and Improved Graphene Synthesis Method

3.1 Exploration of Carbon Sources

Graphene synthesis depends mainly on the carbon source. Different kinds of carbon sources are available today, including graphite, C_{60} fullerene, amorphous carbon (amorphous carbon), etc. Among which, the most common type is graphite, which is a crystalline form of carbon, ample in nature.

Graphite can be used to produce graphene in the form of graphite flakes using the sonication method and the graphite oxide reduction method. If it is used as a solid graphite block, it can be utilized in the exfoliation method. However, the scalable production of large sheets of graphene using this source is still a challenge due to the defects and cracks in graphite [1].

C_{60} fullerene molecules can also be used to produce graphene due to their similar structure. C_{60} is the most abundant among fullerenes, available at relatively low cost, stable at room temperature, decomposition occurs at relatively lower temperature range, i.e., 700–1000°C [2]. However, graphene wrinkles crossing the substrate terraces make it unfit for mass production [3].

Graphene growth using amorphous carbon as a carbon source requires high temperature and high pressure [4]. These conditions are required to rearrange the atomic structure of carbon source,

---

Graphene and VLSI Interconnects
Cher-Ming Tan, Udit Narula, and Vivek Sangwan
Copyright © 2022 Jenny Stanford Publishing Pte. Ltd.
ISBN 978-981-4877-82-4 (Hardcover), 978-1-003-22488-4 (eBook)
www.jennystanford.com

which is 3D in nature initially, to the final product that is 2D in nature, and thus high energy is required to break the chemical bonds for such transformation. However, better quality and large sheets of graphene are feasible with this carbon source, and this is the main motivation of this work to use amorphous carbon in the production of graphene at relatively larger scale.

There are reports in the literature where amorphous carbon is used as the solid carbon source to obtain non-transferable graphene on nickel or cobalt thin film, which is used as both a substrate and catalyst, and annealed at elevated temperatures [5, 6]. This proposed technique can be very helpful in order to:

- Engineer the graphene layers,
- Eliminate the need for graphene transfer, and
- Be economical because gas sources are replaced by a solid carbon source.

However, such a synthesis method using copper as catalyst was adjudicated as infeasible by some research groups [5, 7].

3.2 Amorphous Carbon as a Promising Candidate: A PVD-Based Synthesis

As can be seen earlier, the common chemical vapor deposition (CVD) method for graphene synthesis is not promising and viable for very large-scale integration (VLSI) interconnects, and there is a need for a new graphene growth method. A novel physical vapor deposition (PVD)-based method has thus been developed for graphene on copper. There are different placements of carbon source, and the most suitable placement that facilitates graphene synthesis will be discussed.

3.2.1 Exploration of Different Placements of Carbon Source

3.2.1.1 Sample preparation using PVD method

Different placements of the carbon source were prepared with this novel PVD method, as shown in Fig. 3.1. In all the placements, the amorphous carbon thin film is of 60 nm, and the copper (99.99%) film is of 800 nm over the Si/SiO$_2$ substrate whose thickness is 300 nm. Samples with different placements of amorphous carbon are labeled as S1 (amorphous carbon layer beneath the copper thin film), S2 (amorphous carbon layer on the top of copper thin film), S3 (copper thin film sandwiched between amorphous carbon layers), and S4 (amorphous carbon layer sandwiched between copper thin films).

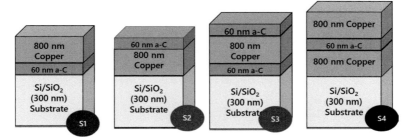

Figure 3.1 Different placements of amorphous carbon source: (a) sample S1, 60 nm amorphous carbon beneath the 800 nm copper thin film; (b) sample S2, 60 nm amorphous carbon layer on the top of the 800 nm copper thin film; (c) sample S3, 800 nm copper thin film sandwiched between the two 60 nm amorphous carbon layers; (d) sample S4, 60 nm amorphous carbon is sandwiched between the 800 nm copper thin films. Reprinted from Ref. [8], Copyright 2016, with permission from Elsevier.

Amorphous carbon and copper deposition were performed using RF and DC sputtering, as shown in Figs. 3.2–3.4. Samples labeled as S1.1 and S1.2 are variants of sample S1, as shown in Fig. 3.2.

3.2.1.2 Post-PVD annealing

Subsequent annealing of all the samples at a temperature of 1020°C was performed for 50 min, in hydrogen (99.99% pure H$_2$) environment with a flow rate of 50 sccm at a low pressure of 1 Torr after the sample preparation. The samples were then cooled down, during which the H$_2$ flow rate is decreased to 30 sccm and argon (Ar) gas is introduced at a flow rate of 20 sccm. A schematic of the annealing chamber and recipe used for annealing is shown in Figs. 3.5 and 3.6, respectively.

30 | *Novel and Improved Graphene Synthesis Method*

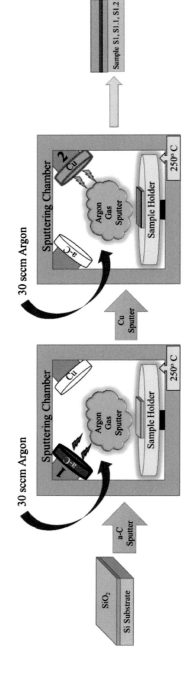

- Copper Thickness – 800 nm
- a-C Thickness – 60 nm for sample S1, 36 nm for sample S1.1 and 12 nm for sample S1.2

Figure 3.2 Sample S1, S1.1, and S1.2 preparation process using PVD.

Amorphous Carbon as a Promising Candidate: A PVD-Based Synthesis | 31

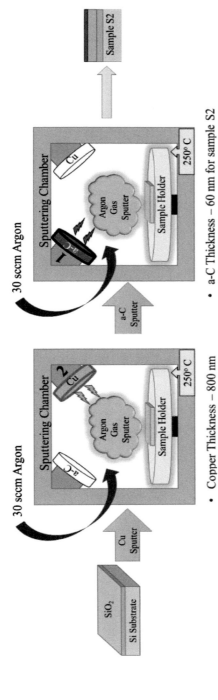

Figure 3.3 Sample S2 preparation process using PVD.

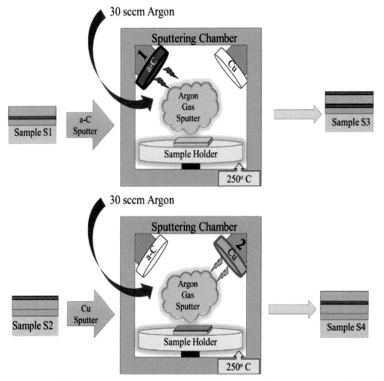

Figure 3.4 Sample S3 and S4 preparation process using PVD.

3.2.1.3 Characterization

The PTT RAMaker Micro Raman/PL/TRPL Spectrometer equipment was employed for the characterization of the annealed samples, which has a confocal Raman microscope system possessing a laser wavelength of 473 nm with the spot size of 0.5 µm approximately.

To deduce the presence of graphene over the substrate, three signature peaks, namely, I_D, I_G, and I_{2D}, were identified and analyzed from the Raman characterization [9]. All these three peaks also play a crucial role for defects detection in graphene. The I_G peak corresponds to E_{2g} (doubly degenerate) phonon at the Brillouin zone center, which originated from the sp^2-bonded C–C stretching vibrations and expressed by the first order of Raman scattering

[10]. The I_D peak is associated with the iTO phonons around the K point of the Brillouin zone and occurred due to the double resonance phenomena. This peak helps to identify the defects in the sample. I_{2D} is associated with the second-order zone boundary that involves two iTO phonons, and a single strong peak appeared for the monolayer graphene. In other words, it helps to identify the number of graphene layers and the grain size [10].

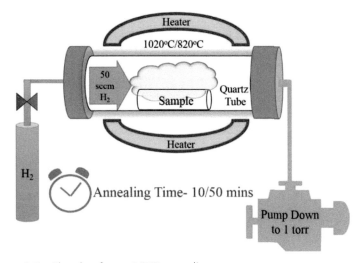

Figure 3.5 Chamber for post-PVD annealing.

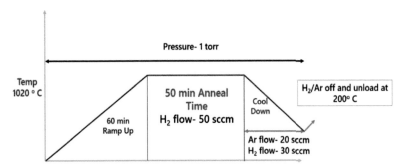

Figure 3.6 Recipe for post-PVD annealing.

To discern the number of layers, i.e., whether graphene is monolayer or bilayer, or more, the I_{2D}/I_G peak intensity ratio was used. If the peak ratio is 1 or greater than 1, it represents that few layers (i.e., layer number around 3–9) are present, whereas if the peak ratio is less than 1, it signifies the presence of multilayer graphene (i.e., layer number more than 10) [11].

To identify the quality of graphene or to determine the defects, the I_D/I_G peak intensity ratio was utilized [10–12]. The values of the I_D/I_G peak intensity ratio closer to 1 suggest that the level of defects is high, which becomes evident with not only the intensity values in particular, but also with the concurrent growth in the full width at half maximum (FWHM) values as well. Generally, the I_D/I_G peak intensity ratio is a good way to understand the level of defects in the grown graphene film.

The Raman spectrum of post-PVD annealed sample S1 is shown in Fig. 3.7, and it can be clearly observed that multilayer graphene is present with the I_D/I_G and I_{2D}/I_G peak intensity ratios of 0.49 and 0.29, respectively. From this, it is evident that the multilayer graphene is feasible from the crystallization of amorphous carbon, catalyzed by copper.

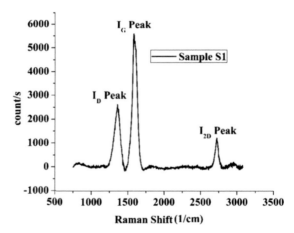

Figure 3.7 Raman spectrum of post-PVD annealed sample S1. Reprinted from Ref. [8], Copyright 2016, with permission from Elsevier.

The Raman spectra of annealed samples S2, S3, and S4 are shown in Fig. 3.8. It can be clearly seen that no significant I_{2D} peak is present in S4, whereas in samples S2 and S3, substantial but distinctive I_G and I_D peaks exist.

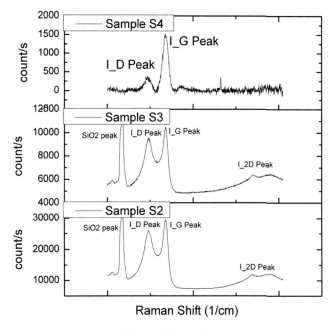

Figure 3.8 Raman spectra of annealed samples S2, S3, and S4. Reprinted from Ref. [8], Copyright 2016, with permission from Elsevier.

In sample S4, amorphous carbon is sandwiched between the copper layers and these copper layers provide a catalysis site for the commencement of carbon crystallization, and graphitization of amorphous carbon is more likely. Consequently, rapid crystallization occurred, but when the annealing time ends (which is 50 min), almost all graphene etched off via H_2 gas and results in disordered graphite [15]. This happens due to double copper layers in a sandwich fashion, which facilitates graphitization rapidly causing large amounts of hydrogen radical creation, which shifts the crystallization reaction equilibrium and results in more graphene etching.

The Raman spectra of samples S2 and S3 resemble with either glassy carbon [13] or graphene oxide [14].

As the Raman spectrum of sample S1 showed promising results, it is further explored in order to understand the underlying mechanism for such growth possibility. The following experimentation was designed with sample S1 configuration as the main focus.

3.2.2 In-Depth Analysis of Best Placement of Carbon Source for Graphene Synthesis

3.2.2.1 Sample preparation using PVD method

Various thicknesses of amorphous carbon thin films were deposited on Si/SiO$_2$ (300 nm) substrates followed by a copper (99.99%) film of 800 nm, as shown in Fig. 3.3. Each of the samples with different amorphous carbon layer thickness is labeled as sample S1.1 (amorphous carbon layer thickness of 36 nm) and sample S1.2 (amorphous carbon layer thickness of 12 nm). An important piece of information extracted from the preparation of samples is the crystal orientation of the copper layer, which is found to be {111} from the XRD analysis, as shown in Fig. 3.9. The {111} crystal orientation of copper is considered to be favorable for high-quality single-crystal graphene synthesis [16] as there is a minimum lattice mismatch between {111} copper and graphene [17].

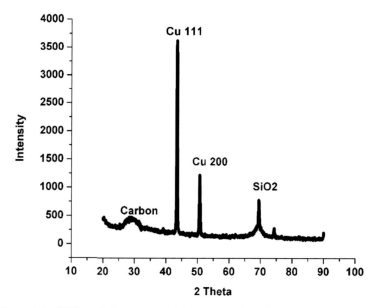

Figure 3.9 XRD analysis on deposited copper surface shows the presence of a high-intensity {111} copper peak in the XRD analysis result.

3.2.2.2 Post-PVD annealing and characterization

All the samples are subsequently annealed with the same recipe, as shown in Fig. 3.6.

The Raman spectra of the annealed samples S1.1 and S1.2 are shown in Figs. 3.10 and 3.11, respectively, and the following information can be obtained. For samples S1.1, the I_{2D}/I_G intensity peak ratio is much higher (close to 1) in comparison to S1, while the I_D/I_G peak ratio is nearly the same as S1. However, for sample S1.2, the I_{2D}/I_G peak ratio is also higher (but less than 1), and its I_D/I_G peak ratio is relatively much less than sample S1. This information is summarized in Table 3.1.

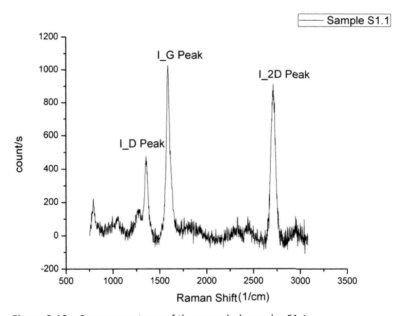

Figure 3.10 Raman spectrum of the annealed samples S1.1.

Table 3.1 clearly depicts that all three samples, i.e., S1, S1.1, and S1.2, have the I_{2D}/I_G peak ratio less than 1, which indicates that they are multilayer graphene. Among all these samples, the maximum number of layers is obtained in sample S1, while the least number of layers is found in sample S1.1, even though the thickness of amorphous carbon in S1.1 is not the least. This indicates that there is optimum amorphous carbon thickness in the fabrication process in order to obtain good-quality graphene.

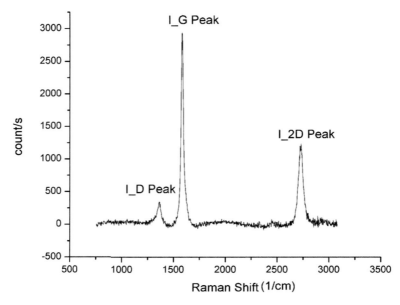

Figure 3.11 Raman spectrum of the annealed samples S1.2.

Table 3.1 Raman characteristics for annealed samples S1, S1.1, and S1.2 at 1020°C for 50 min

Sample	I_D Peak Intensity [cm^{-1}]	I_G Peak Intensity [cm^{-1}]	I_{2D} Peak Intensity [cm^{-1}]	I_D/I_G Ratio	I_{2D}/I_G Ratio	2D Peak FWHM Value [cm^{-1}]
S1 (with 60 nm amorphous carbon)	1362	1584	2726	0.496	0.29	48
S1.1 (with 36 nm amorphous carbon)	1355	1589	2711	0.464	0.89	56
S1.2 (with 12 nm amorphous carbon)	1367	1584	2730	0.117	0.42	52

Source: Reprinted from Ref. [8], Copyright 2016, with permission from Elsevier.

Amorphous Carbon as a Promising Candidate: A PVD-Based Synthesis | 39

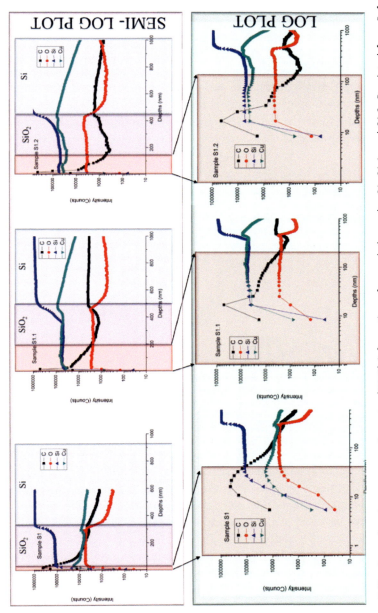

Figure 3.12 Secondary ion mass spectrometry (SIMS) information of annealed samples S1, S1.1, and S1.2. Reprinted from Ref. [8], Copyright 2016, with permission from Elsevier.

The secondary ion mass spectrometry (SIMS) was employed to analyze the composition of graphene thin films that are produced over the Si/SiO$_2$ substrate via the PVD sputtering method. These SIMS results are taken after annealing of the samples (i.e., S1, S1.1, and S1.2) for 50 min, and they are shown in Fig. 3.12. In this technique, a focused primary ion beam is projected over the area of interest, and scattered secondary ions were collected. The mass spectrometer analyzes the ratio of mass and charge of the secondary ions that helps to determine the molecular composition of the samples from the surface to depth.

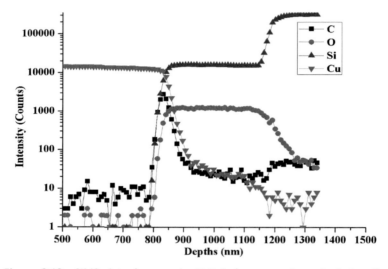

Figure 3.13 SIMS data for sample S1.2 before annealing, displaying the presence of amorphous carbon layer beneath the copper film.

For all the samples, the oxygen content is almost negligible. Large and wide intensity peaks of carbon were easily identifiable in sample S1, which represents that carbon is present at the depths, signifies multiple layers of graphene exist in sample S1. This reinforced our Raman characterization results, which also identified the presence of multiple graphene layers. Similarly, in samples S1.1, a sharp and thin carbon intensity peak was detected, which represents the existence of carbon over a thin layer in comparison to S1 and signifies the presence of few layers of graphene. In sample S1.2, a wider (as compared to S1.1) but sharper (as compared to S1) peak is found because carbon atoms are situated at smaller depths in comparison to sample S1. Therefore, the number of graphene layers in S1.2 is found in the middle of samples S1 and S1.1, in agreement with the Raman characterization discussed earlier.

When Fig. 3.12 is compared to Fig. 3.13, which shows a log-plot of SIMS data for sample S1.2 before annealing (but with amorphous carbon), it is seen that there is carbon diffusion into SiO_2 and Si, through copper and exist on the surface of copper. The carbon on the surface of copper is likely to be graphene as confirmed by the Raman characterization.

3.3 Growth Mechanism of PVD-Based Graphene

The underlying mechanism of this novel PVD process for the growth of graphene on copper can be understood by the following studies:

1. Study of the effect of amorphous carbon thickness
2. Study of the effect of annealing time
3. Study of the effect of annealing temperature

Finite element modeling (FEM) simulations are also designed in order to establish a comprehensive understanding of the mechanism.

3.3.1 Effect of Amorphous Carbon Thickness

Figure 3.14 summarizes the study of PVD-based graphene growth with respect to variable thickness, as discussed in Section 3.2.

Figure 3.14a shows the configuration of an experimental sample, where the amorphous carbon layer (variable thickness of 60 nm, 36 nm, 12 nm) sitting under the 800 nm thick copper film. This sample is prepared by the PVD method (RF and DC sputtering) as explained in the previous section. Figure 3.14b shows the post-PVD annealing chamber with annealing conditions and parameters.

Figures 3.14c–e show the Raman spectra for the samples S1, S1.1, and S1.2 with three different thicknesses 60 nm, 36 nm, and 12 nm, respectively. As the I_{2D}/I_G intensity peak ratio is less than 1 for all the three samples, it shows that the obtained graphene is multilayer, where sample S1 possesses the maximum number of layers and

42 | *Novel and Improved Graphene Synthesis Method*

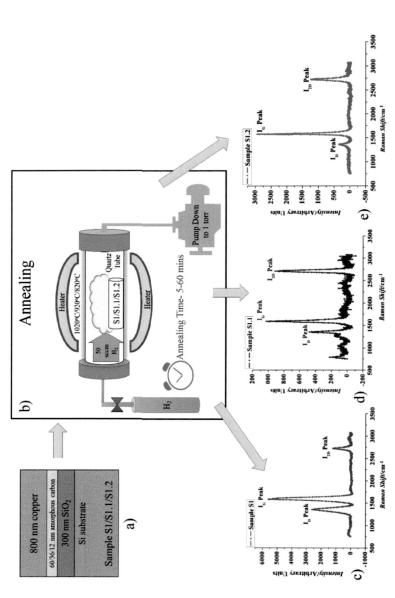

Figure 3.14 (a) Structure design for samples S1, S1.1, and S1.2 having 60 nm, 36 nm, and 12 nm amorphous carbon layer sandwiched between 800 nm copper and Si/SiO$_2$ (300 nm) substrate, (b) schematic for annealing process, (c) Raman spectrum of annealed sample S1, (d) Raman spectrum of annealed sample S1.1, (e) Raman spectrum of annealed sample S1.2 [18].

sample S1.1 possesses the minimum number of layers, even though the amorphous carbon thickness in sample S1.1 is not the least. The defect density indicated by the I_D/I_G peak ratio for samples S1 and S1.1 is nearly the same, while for sample S1.2, it is much smaller.

3.3.2 Effect of Annealing Time

Samples S1.1 and S1.2 were exposed to different annealing times, and higher 2D peaks were observed in the Raman spectrum, as shown in Fig. 3.15. The Raman mappings of samples S1.1 and S1.2 at different annealing times but at constant temperature, i.e., 1020°C, are shown in Figs. 3.16 and 3.17.

The means plotted with error bars in Fig. 3.15 are consistent with the Raman mappings in Figs. 3.16 and 3.17.

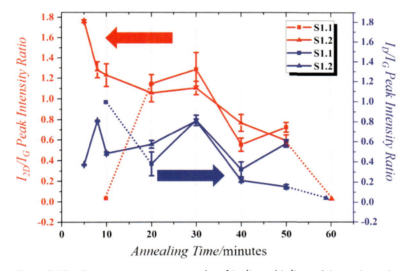

Figure 3.15 Raman spectroscopy results of I_{2D}/I_G and I_D/I_G peak intensity ratio versus annealing time (minutes). Dotted lines indicate the uncertainty of the trend as mentioned in the text [18].

The I_{2D}/I_G peak intensity ratio is an indicator of the presence of graphene. In sample S1.1, for the first 10 min, no graphene layer can be observed as its surface is highly defective, which is indicated by its I_D/I_G ratio. After 20 min of annealing time, the I_D/I_G ratio drops to lower value, whereas the I_{2D}/I_G ratio increases, showing the formation of graphene layers has been initiated. The maximum I_{2D}/I_G ratio (greater than 1) is achieved after 30 min of annealing time, which shows the presence of a few layers of graphene layers only; however, the I_D/I_G ratio increases at the same time. A subsequent decrease in the ratios of I_D/I_G and I_{2D}/I_G can be observed at 40 min of annealing time, and at 50 min of annealing time, a slight increase in both the ratios can be perceived. In these experiments, the discrete annealing time is taken, which is not appropriate to detect the sharp change in the ratios (these sharp changes are anticipated by the dotted lines in Fig. 3.15).

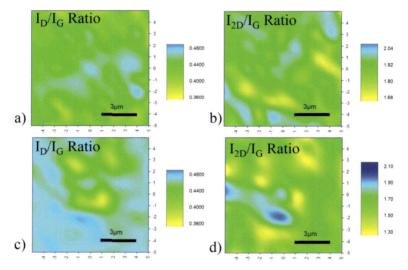

Figure 3.16 Raman mapping of sample S1.1 (10 µm × 10 µm) area with the step size of 1 µm, which is annealed at 1020°C temperature for (a, b) 20 min and (c, d) 30 min [18].

While sample S1.2 goes through the early crystallization of a thin amorphous carbon layer within the first 5 min of annealing time and results in the generation of a few layers of graphene, in the next 25 min, multilayered graphene is obtained with a trivial increase in the I_{2D}/I_G ratio. In the following annealing time intervals, a decrease in the ratios of I_{2D}/I_G and I_D/I_G intensity peaks occurred in comparison to sample S1.1. In contrast, the defects in sample S1.2 are relatively less than sample S1.1 after 40 min of annealing time.

Growth Mechanism of PVD-Based Graphene | 45

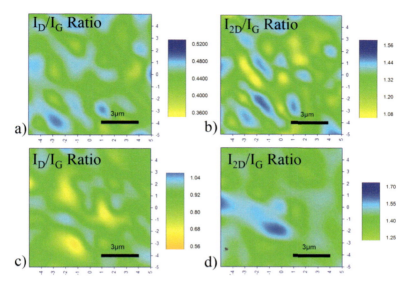

Figure 3.17 Raman mapping of sample S1.2 (10 μm × 10 μm) area with the step size of 1 μm, which is annealed at 1020°C temperature for (a, b) 5 min and (c, d) 8 min [18].

This finding helps to identify the reasons behind the infeasibility of the crystallization of amorphous carbon using copper as a catalyst in the previously reported work as the authors were either using annealing time less than 10 min, which is not enough, or the amorphous carbon is too thick that the creation of graphene will only be possible at longer annealing times [7].

3.3.3 Effect of Annealing Temperature

Samples S1.1 and S1.2 are subjected at comparatively lower annealing temperatures, i.e., 920°C and 820°C, for the duration of 50 min in order to determine the effect of temperature on graphene growth and their respective Raman spectra as presented in Fig. 3.18.

The presence of high 2D peaks at a relatively lower temperature regime indicates a possibility of graphene growth at such a low temperature. At 820°C, graphene can be obtained in sample S1.2 in contrast to sample S1.1. However, if the annealing temperature is further reduced to 800°C, no graphene can be observed in both samples.

46 | Novel and Improved Graphene Synthesis Method

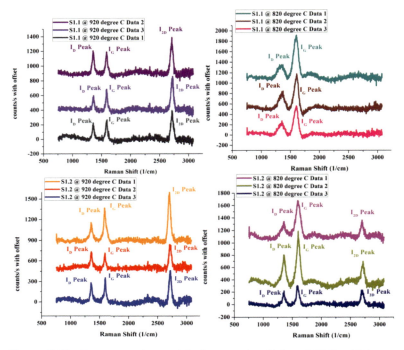

Figure 3.18 Raman spectrum plots for samples S1.1 and S1.2, which are annealed at 920°C and 820°C for 50 min [18].

A clear difference in the behavior of graphene formation can be observed in samples S1.1 and S1.2 when they were subjected at different annealing times and temperatures, which is very interesting, as the only difference between these two samples is the amorphous carbon layer thickness. This peculiar behavior is assumed to have occurred due to the thermomechanical stress-driven crystallization of amorphous carbon into graphene, and it is verified using finite element analysis (FEA).

3.3.4 Stress Analysis Using Finite Element Modeling

ANSYS Workbench R16.2 was employed to perform FEA in order to perform the stress analysis of graphene over the substrate.

Thermomechanical stress [19] is given as:

$$\sigma_{sth} = \left[\frac{E_f}{(1-v_f)}\right]*(\alpha_s - \alpha_f)*\Delta T \qquad (3.1)$$

where E_f is Young's modulus, v_f is Poisson's ratio, α is coefficient of thermal expansion, and ΔT is the range of temperature expansion. Here, subscripts "s" and "f" refer to substrate and film, respectively.

It can be clearly seen in Eq. 3.1 that the amorphous carbon thickness has no impact over the thermomechanical stress. Therefore, each sample should experience the same thermomechanical stresses under the graphene growth process. However, this thermomechanical stress causes the sample to warp, and this results in maximum principal stress given by Stoney's equation [20, 21] as follows:

$$\sigma_{stf} = \left[\frac{E_f}{6(1-v_f)}\right] * \left(\frac{t_s 2}{t_f}\right) * \Delta\left(\frac{1}{R}\right) \tag{3.2}$$

where $\Delta(1/R) = (1/R - 1/R_0)$, R_0 is the radius of curvature of the samples before the thermomechanical stresses are developed, R is the radius of curvature of the samples after the thermomechanical stresses are developed, and t is the thickness of the layer in the sample for the film and substrate.

Since amorphous carbon is considered a brittle material [22], the maximum principal stress is considered through Stoney's equation in this case [22].

In Eq. 3.2, the $\Delta(1/R)$ term is because of the warpage of the sample due to the thermomechanical stress, and it does not depend upon the thickness of the layer. Therefore, it is assumed that it will be constant for all the three samples.

With all these factors taken into consideration, the 3D simulation model is developed to analyze the thermomechanical stress where each material property is provided in Table 3.2.

Table 3.2 Thermomechanical properties for stress analysis using FEM

Property	SiO$_2$ Film	Amorphous Carbon Film	Copper Film	Graphene
Thermal coefficient of expansion [°C^{-1}]	5.0×10^{-7} [23]	1.5×10^{-6} [24]	1.6×10^{-5} [25]	-8×10^{-6} [26]
Young's modulus [TPa]	0.070 [27]	0.759 [28]	0.115 [29]	0.96 [30]
Poisson's ratio	0.1700 [27]	0.1700 [28]	0.3430 [29]	0.1700 [30]

48 | *Novel and Improved Graphene Synthesis Method*

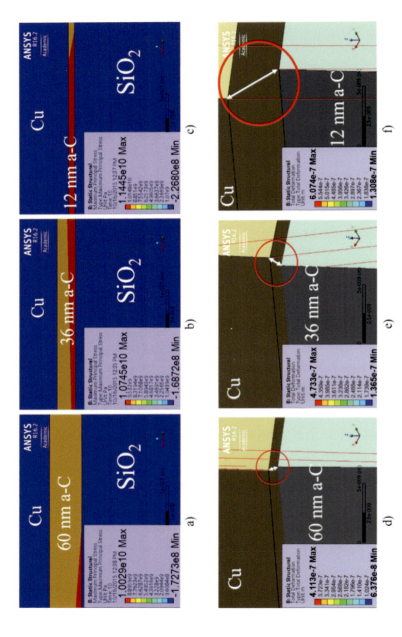

Figure 3.19 ANSYS® simulation results of (a–c) maximum principal stress distributions in the three samples at 1020°C, zooming into the maximum stress areas; (d–f) von Mises stress in copper at the corner of each sample, showing the delamination at the amorphous carbon/substrate and copper/amorphous carbon interfaces in these samples [18].

Stoney's equation (Eq. 3.2) shows that the maximum principal stress is inversely proportional to the thickness of the amorphous carbon in our case. In our experiments and simulation models, the thicknesses of the samples S1, S1.1, and S1.2 are 60 nm, 36 nm, and 12 nm, respectively. Therefore, the minimum stress should be experienced by sample S1 and the maximum stress by sample S1.2. The same results are supported by our simulation results, as shown in Figs. 3.19a–c at 1020°C.

Figures 3.19d–f are the zoomed-in images of Figs. 3.19a–c where the main focus is at the interface of copper and amorphous carbon. As copper is ductile [31] and amorphous carbon is brittle in nature, the equivalent von Mises stress is only considered at copper. It can be clearly observed in the images that the delamination of copper and amorphous carbon films occurred at the edges where the maximum stress is found. Most severe delamination is observed in sample S1.2 because of the presence of maximum stress in copper and amorphous carbon layers as predicted earlier. The copper film is under tensile stress, so it is believed that when high von Mises stresses are present, it can either crack the copper film and/or enhance the hydrogen gas diffusion in the copper film (through its grain boundaries) [32, 33].

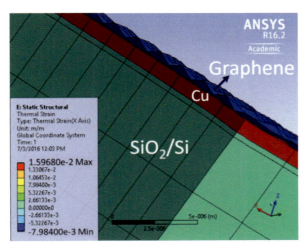

Figure 3.20 Computed thermal strain in graphene on copper thin film at 1020°C in ANSYS® [18].

Our thermal strain analysis of the synthesized graphene on the copper using the ANSYS simulation tool, as shown in Fig. 3.20, shows that the copper thin film has tensile strain (positive value), while the graphene layer has compressive strain (negative strain).

Based on FEA simulation results, the experimental results are revisited and explained in the following discussion.

3.3.5 Interpretation of Experimental Results with the Aid of FEM

The number of graphene layers depends on the initial thickness of amorphous carbon [5], and this is true for sample S1, as it possesses the highest amorphous carbon thickness, i.e., 60 nm, as verified in Fig. 3.14c. However, it is not true for sample S1.2, as it should have the minimum layers of graphene because its thickness is 12 nm only. The minimum number of graphene layers was found in sample S1.1, whose thickness is 36 nm, more than that of sample S1.2, as shown in Fig. 3.14.

To understand this abnormality, the well-founded chemical reactions that govern the graphene growth, as given by Vlassiouk et al. [34], are employed, in addition to our I_{2D}/I_G and I_D/I_G intensity peak ratios (as shown in Fig. 3.15) for further analysis. These chemical reactions are shown in Fig. 3.21, but they are modified for our experiments since methane is not used. Here, reaction (1) occurs between the amorphous carbon and H_2 gas at elevated temperatures, which produces the hydrocarbons as it is also reported by Ji et al. [35].

For sample S1.1, the obtained graphene is totally defective when the annealing time is below 10 min, as shown by the close-to-zero I_{2D}/I_G intensity peak and very high I_D/I_G intensity peak in Fig. 3.15. It happened because of the lower tensile stress in the copper film; therefore, hydrogen requires more time to reach the underneath amorphous carbon for graphene crystallization.

Graphene crystallization increases as the annealing time increased to 20 min. This rate of increase is as per the endothermic reaction (2) as mentioned in Fig. 3.21 [32]. Meanwhile, hydrogen gas performs dual functions in this reaction: as an activator and as a graphene etchant. This gas helps in the crystallization of graphene (increase the number of layers) as an activator, whereas as an etchant, it etches away the already grown graphene layers. This dual role results in the reduced number of graphene layers as verified by our Raman spectrum results in Fig. 3.14, where the value of the I_{2D}/I_G ratio is increased. However, as the annealing time is increased to 20 min from 10 min, better quality graphene is produced as indicated by the decreased value of I_D/I_G. This is due to the availability of hydrogen radicals in reaction (3) (which is an exothermic etching reaction) that shifts the reaction to go in the forward direction [36], which enables the production of carbon radicals according to reaction (1). However, discrete annealing time intervals used here in our experiments make it difficult to determine the crystallization and etching of graphene during the time interval between 10 and 20 min.

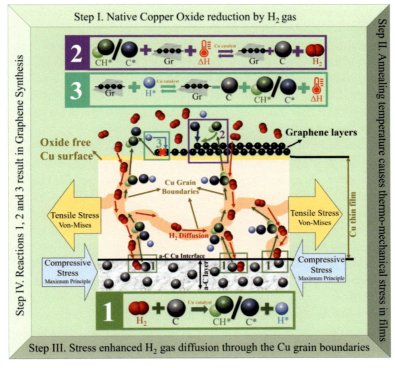

Figure 3.21 Governing reactions [9, 34] for graphene growth forming the basis for the proposed mechanism of graphene synthesis on copper film; here C represents a carbon atom, Gr represents graphene, CH*/C* represent hydrogen-bonded carbon radical/carbon radical, H* represents hydrogen radical, and ΔH represents heat [18].

The number of hydrogen radicals becomes sufficient at 30 min of annealing time to initiate reaction (3). Likewise, the heat of absorption owing to reactions (1) and (2) decreases the temperature of the copper film surface. Consequently, the graphene crystallization rate subjugated the graphene etching rate. Henceforth, graphene etching primarily occurs at the defective sites as they facilitate as the preferential sites of etching due to the availability of H_2 [37]. As a result, a few graphene layers with increased defective sites are obtained as compared to former shorter annealing times. Therefore, the I_{2D}/I_G and I_D/I_G intensity peak ratios show increment in their magnitudes at the annealing time of 30 min, as shown in Fig. 3.15.

At 40 min of annealing time, more graphene layers can be observed as the I_{2D}/I_G ratio reduced to a value less than 1, as shown in Fig. 3.15. This annealing period provides a sufficient amount of time to generate the heat from reaction (3) and as a consequence, the amount of by-products (carbon radicals) increases and enhances the graphene crystallization process (as reaction (2) operates in the forward direction). Additionally, decrement in the I_D/I_G ratio is also detected as it is believed that the rate of newly formed graphene layers is more than the etching rate. Hence, new graphene layers fill the defective sites and thus good quality of graphene is achieved.

In 50 min of annealing time, the same trend (of graphene formation) is observed as that during 20 to 30 min of annealing time, i.e., the formation of hydrogen radicals with carbon radicals. This trend increases the number of hydrogen radicals in the chamber, which reduces the temperature of the copper thin film surface. Reactions (1) and (2) activate reaction (3) due to the heat of absorption, and as a consequence, it shifts to forward direction. Hence, fewer graphene layers are obtained and verified with the increase in the I_{2D}/I_G intensity peak ratios, as shown in Fig. 3.15. Subsequently, an increase in the I_D/I_G intensity peak ratio can also be observed in Fig. 3.15 because of the increase in the defects of the layers due to the loss of carbon atoms from the graphene crystal, forming defective sites. This action is substantiated from the Raman spectrum results of sample S1.2 in Fig. 3.14e. This phenomenon can be possible due to the high tensile stress in copper, which increases the hydrogen rate of diffusion through the copper film, and graphene synthesis is observed with low I_D/I_G intensity peak ratio for as low as 5 min of annealing time, as shown in Fig. 3.15. Sample S1.2 has higher tensile stress, and thus the carbon radical supply on the surface of copper is more due to the enhanced H_2 diffusion through the copper thin film. This provides a favorable condition for reaction (2), which keeps enabling the continued growth of graphene.

The annealing time between 8 and 20 min has shown an increase in the layer numbers of graphene as presented by the decrease in the I_{2D}/I_G intensity peak ratio in Fig. 3.15. Correspondingly, a decrease in the I_D/I_G intensity peak ratio can also be observed as the newly formed graphene layers fill the defective sites, resulting in a reduction in the number of defective sites that corresponds to a production of good-quality graphene. More number of hydrogen radicals are produced due to the high rate of reaction (1), and the collective effect of reactions (1) and (2) causes the heat of absorption, which reduces the copper thin film surface temperature. The cooperative effect of these two factors renders reaction (3) in the forward direction, and as a result, graphene layers start reducing in numbers if the annealing time exceeds 30 min. Hence, after 30 min of annealing time, the number of graphene layers does not increase continuously; instead, the reduction in the number of layers starts slowly due to the dominance of reaction (3). This is verified by a small increase in the I_{2D}/I_G intensity peak ratios, whereas an increase in the I_D/I_G intensity peak ratio can also be observed due to etching, as elucidated earlier.

The subsequent decrease in the I_{2D}/I_G and I_D/I_G intensity peak ratios can be observed after 30 min of annealing time as reaction (2) comes into effect. The carbon radicals produced as a by-product in the previous stage, owing to reaction (3), shift the graphene growth reaction (2) in the forward reaction and provide lesser chance to hydrogen radicals to etch away the graphene. However, if the annealing time is prolonged (40 min and beyond), then a different type of mechanism kicks in, which is unlikely to be the same in sample S1.1. This can be seen (in Fig. 3.15) from the nonuniform rate of change in the I_{2D}/I_G ratio, which follows a different trend as that of sample S1.1. This is believed to be due to two possible mechanisms: stress-induced migration [38] and a sharp increase in the amorphous carbon film compressive stress.

Stress-induced migration or stress-induced voiding [39, 40] increases the number of vacancies/voids as well as their movement in the copper thin films. These vacancies move toward the edges of the copper thin films and combine with other vacancies and result in bigger vacancies. These vacancies work as active sites for the diffusion of hydrogen and carbon radicals. This postulation is made on the basis of the computed tensile stresses in the ANSYS simulation tool as described earlier in the Cu film on SiO_2 substrate with and without a C film. Finite element calculations using Eqs. (3.1) and (3.2) show that the tensile stress in the copper film for the former is 1190 MPa and that for the latter ranges from 460 MPa to 690 MPa depending upon the amorphous carbon layer thickness, as depicted in Fig. 3.22.

The amorphous carbon film thickness was 12 nm for sample S1.2 initially. After 40 min of annealing time, this thickness reduced to a very thin layer. Consequently, the tensile stress in the copper film becomes high, which can initiate the stress-induced voiding process in the sample, and this effect is more severe at a higher temperature through the annealing process. This hypothesis is strongly supported by the evidence that is obtained from the ANSYS simulations, as shown in Fig. 3.23.

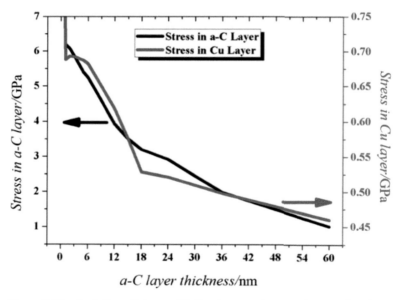

Figure 3.22 Evolution of stress with the decrease in amorphous carbon layer thickness [18].

Growth Mechanism of PVD-Based Graphene | **55**

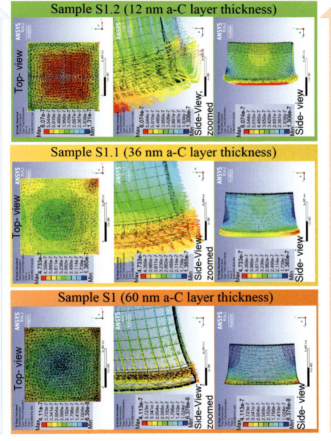

Figure 3.23 Total deformation results obtained from ANSYS® for samples S1, S1.1, and S1.2. This shows the migration of deformation from the corners to the center while the thickness of amorphous carbon decreases [18].

Figure 3.23 shows that the maximum deformation area in samples S1 and S1.1 occurs at the corners, whereas in the case of sample S1.2, it is densely concentrated at the center. The reason for this is that when the thickness of amorphous carbon is high initially and, at this moment, corners are comparatively free to deform, high stress values are found at the corners. With the decrease in the thickness of amorphous carbon during the process, the stress at the interface of amorphous carbon and copper increases slowly, which causes delamination at the corners of the samples, as depicted in Figs. 3.19d,e. Therefore, corners become stress free in the initial stage, and later this deformation migrates to the center area causing the higher stress distribution at the center. Hence, the free carbon radicals come out slowly from the corners of samples S1 and S1.1. On the other hand, with the reduced thickness of amorphous carbon, the movements of these free radicals are more from the center as compared to their counterpart. In other words, the motion of carbon radicals initiates from the center for sample S1.2, and over time, this distribution area keeps widening and reaches the edges of the corners and delamination occurs.

The second mechanism, i.e., a sharp increase in the amorphous carbon film compressive stress, is found from the computation of FEM simulation using ANSYS, as shown in Fig. 3.19. This mechanism is found in sample S1.2 where the initial thickness of amorphous carbon was 12 nm; however, if the annealing times exceed the 40 min duration, this thickness reduced to 4 nm or below. Therefore, the resultant stress becomes very high in the amorphous carbon layer, as shown in Fig. 3.22. This high stress "pushes" the carbon atoms from the layers in the perpendicular directions to release the stress, which results in the enhanced diffusion of carbon and hydrogen atoms.

The collective effect of the two aforementioned mechanisms helps in graphene crystallization at a faster rate than its etching. Consequently, a greater number of graphene layers were synthesized in sample S1.2. With this higher rate of diffusion of carbon radicals and the formation of graphene, the initial thin layer of amorphous carbon in S1.2 is depleting quickly, and its remaining amount becomes very small when annealing time goes up to 60 min and beyond. At this time, reaction (2) shifts in the backward direction, which results in the dominance of graphene etching reaction over the deposition reaction, and triggers etching of all the graphene layers by H_2 gas. The exact time when the growth of graphene layers stopped and was overcome by the etching reaction is unknown due to the use of discrete annealing time intervals used in these experiments.

Conversely, the aforementioned mechanisms are unlikely to occur in sample S1.1 because of the following two reasons. First, in the beginning, the thickness of amorphous carbon is 36 nm and after 40 min of annealing time, its thickness is much more than 4 nm. Hence, the compressive stress has not reached up to the point where stress-induced voiding can happen, as shown in Fig. 3.22. With the presence of thick amorphous carbon underneath the copper film, it is very unlikely that the tensile stress can undergo stress-induced voiding.

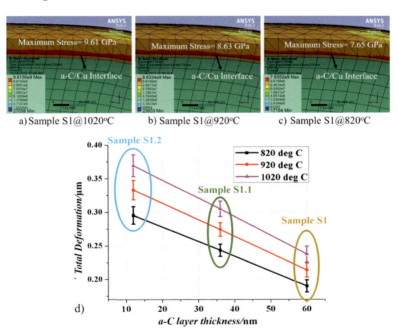

Figure 3.24 (a–c) Simulation results for maximum von Mises stress at a carbon/copper interface for sample S1 with respect to different annealing temperatures, (d) total deformation versus different amorphous carbon thickness plot at different annealing temperatures for samples S1, S1.1, and S1.2 [18].

At a lower annealing time, the magnitude of von Mises stress in copper film and the maximum stressed area in sample S1 are reduced, as depicted by the ANSYS simulation results in Figs. 3.24a–c at different temperatures. As the temperature decreases from 1020°C to 820°C, the maximum stress decreases from 9.61 GPa to 7.65 GPa, respectively.

Therefore, the reaction between the hydrogen gas and amorphous carbon slows down because of the decreased diffusion of hydrogen gas through the copper film. However, the supply of carbon radicals also reduced, but graphene is being etched away due to the presence of the hydrogen gas. Hence, only a single or very few graphene layers are obtained.

If the temperature is further lowered, reaction (1) will advance very slowly because of the slow diffusion of the hydrogen gas through the copper film, which also causes reaction (2) to happen very slowly. Hence, the formation of graphene occurs.

With further reduction in the temperature, the diffusion of the hydrogen gas through the copper thin film will occur at a very slow speed, and because of this, reaction (1) (mentioned in Fig. 3.21) will also progress with slow speed. Consequently, reaction (2) will also occur at a slower speed, and the obtained graphene will be etched away very quickly in the presence of hydrogen gas on top (due to the slow diffusion of hydrogen gas through the copper film, its surface will have comparatively high concentration of hydrogen and thus high etching rate). Therefore, the graphene layer is absent below a certain temperature, which is termed the threshold temperature. Sample S1.2 has the lower threshold temperature because of the thin amorphous carbon film, as compared to sample S1.1, which was verified with our experimental results. The stress in copper for sample S1.2 is maximum, followed by samples S1.1 and S1 at lower temperatures, and this is supported with the total deformation results obtained through ANSYS simulations, as depicted in Fig. 3.24d.

Likewise, the growth mechanism of graphene regarding the annealing temperature value of 1020°C can also be implemented for the lower temperature regimes as the stress follows the same fashion as that of the higher temperature regimes. Figure 3.20 presents the thermal strain results that are verified with the Raman spectroscopy results, as depicted in Fig. 3.25.

Figure 3.25 exhibits the plot of Raman frequencies of G (ωG) versus 2D (ω2D) modes. The purple dashed line with a slope ($\Delta\omega$2D/$\Delta\omega$G) of 2.2 depicts unstrained graphene [41], and point "O" shows intrinsic frequencies of the two modes, which are not affected by strain or excess charges [41, 42]. The chances of doping in the samples are minimum as the experiments are conducted in a vacuum. The Raman frequencies of G and 2D modes for samples S1.1 and S1.2 (represented by red-dotted linearly fitted lines) are nearly parallel to the unstrained graphene line and belong to the compressive strain region [41].

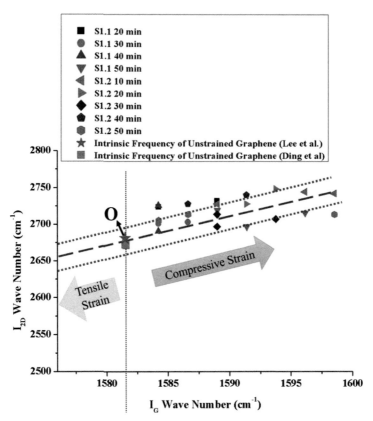

Figure 3.25 The relation between thermomechanical strain and Raman frequencies of G (ωG) and 2D (ω2D) modes; the data points for samples S1.1 and S1.2 (represented by the red-dotted linearly fitted line) lay parallel to the unstrained graphene line [41, 42] (purple dashed line) and indicate compressive strain [18].

3.4 Summary

This chapter shows that graphene synthesis on copper is possible with amorphous carbon as the solid carbon source, and the growth mechanism of graphene is explained. Graphene is successfully synthesized from amorphous carbon sitting under sputtered copper thin film at an annealing temperature of about 1020°C in a pure hydrogen environment. The evidence of graphene is confirmed by its Raman characteristics. This graphene growth method highlights an important aspect of fabrication because the graphene layer designed atop the copper thin film as confirmed from the SIMS data is separated from the amorphous carbon source beneath the copper film. By varying the amorphous carbon thickness, a different number of graphene layers can be formed experimentally, showing the possibility of adjusting the number of graphene layers using the method developed in this chapter.

The ability to synthesize the graphene layer using amorphous carbon source and copper as catalyst enables low-cost application of graphene on copper that can enhance the properties and extend the applications of copper film.

From our experimental analysis, we can observe that the graphene synthesis mechanism on the copper thin film with amorphous carbon source is completely different from the conventional CVD method where a gaseous carbon source is used. This growth mechanism is presented in Fig. 3.21 and summarized as follows:

Step I: Graphene formation process initiates with the reduction of native copper oxide by hydrogen gas at highly elevated temperatures ranging from 820 to 1020°C, which releases water molecules (by-product) during the gas phase and stress-induced grain growth starts [43, 44].

Step II: Thermomechanical stress in the films of samples is developed due to the high annealing temperature, which results in the warping and causing high tensile stress in the copper thin film.

Step III: This high tensile stress in the copper film helps in the diffusion of H_2 gas to the underneath amorphous carbon layer through the copper film and initiates the graphene crystallization.

Step IV: A reaction between the H_2 gas and amorphous carbon occurs, which creates carbon radicals and copper works as a catalyst in this reaction. These carbon radicals move out along the grain boundaries and crystallize to graphene on top of copper. The formation and etching of graphene occur concurrently in the presence of H_2 gas. Different qualities of graphene are achieved in the experiments for different samples, which mainly depends upon three factors: the availability of carbon radicals, amorphous carbon layer thickness, and different stress levels.

References

1. Y. Ma and L. Zhi, "Graphene-based transparent conductive films: Material systems, preparation and applications," *Small Methods*, **3**, 1, pp. 1–32, 2019.
2. F. Cataldo, "A study on the thermal stability to 1000°C of various carbon allotropes and carbonaceous matter both under nitrogen and in air," *Fullerenes Nanotub. Carbon Nanostructures*, **10**, 4, pp. 293–311, 2002.
3. H. Ago, Y. Ogawa, M. Tsuji, S. Mizuno, and H. Hibino, "Catalytic growth of graphene: Toward large-area single-crystalline graphene," *J. Phys. Chem. Lett.*, **3**, 16, pp. 2228–2236, 2012.
4. J. D. Albar et al., "An atomic carbon source for high temperature molecular beam epitaxy of graphene," *Sci. Rep.*, **7**, 1, pp. 1–8, 2017.
5. M. Zheng et al., "Metal-catalyzed crystallization of amorphous carbon to graphene," *Appl. Phys. Lett.*, **96**, 6, pp. 2–5, 2010.
6. G. Pan et al., "Transfer-free growth of graphene on SiO_2 insulator substrate from sputtered carbon and nickel films," *Carbon N. Y.*, **65**, pp. 349–358, 2013.
7. C. M. Orofeo, H. Ago, B. Hu, and M. Tsuji, "Synthesis of large area, homogeneous, single layer graphene films by annealing amorphous carbon on Co and Ni," *Nano Res.*, **4**, 6, pp. 531–540, 2011.
8. U. Narula, C. M. Tan, and C. S. Lai, "Copper induced synthesis of graphene using amorphous carbon," *Microelectron. Reliab.*, **61**, pp. 87–90, 2016.
9. U. Narula and C. M. Tan, "Determining the parameters of importance of a graphene synthesis process using design-of-experiments method," *Appl. Sci.*, **6**, 7, pp. 1–16, 2016.
10. L. M. Malard, M. A. Pimenta, G. Dresselhaus, and M. S. Dresselhaus, "Raman spectroscopy in graphene," *Phys. Rep.*, **473**, 5–6, pp. 51–87, 2009.

11. A. C. Ferrari, "Raman spectroscopy of graphene and graphite: Disorder, electron-phonon coupling, doping and nonadiabatic effects," *Solid State Commun.*, **143**, 1–2, pp. 47–57, 2007.

12. A. C. Ferrari et al., "Raman spectrum of graphene and graphene layers," *Phys. Rev. Lett.*, **97**, 18, pp. 1–4, 2006.

13. D. S. Kozak, R. A. Sergiienko, E. Shibata, A. Iizuka, and T. Nakamura, "Non-electrolytic synthesis of copper oxide/carbon nanocomposite by surface plasma in super-dehydrated ethanol," *Sci. Rep.*, **6**, pp. 1–9, 2016.

14. Y.-H. Lin, C.-Y. Yang, S.-F. Lin, and G.-R. Lin, "Triturating versatile carbon materials as saturable absorptive nano powders for ultrafast pulsating of erbium-doped fiber lasers," *Opt. Mater. Express*, **5**, 2, p. 236, 2015.

15. D. A. Boyd et al., "Single-step deposition of high-mobility graphene at reduced temperatures," *Nat. Commun.*, **6**, pp. 1–8, 2015.

16. M. Ishihara, Y. Koga, J. Kim, K. Tsugawa, and M. Hasegawa, "Direct evidence of advantage of Cu(111) for graphene synthesis by using Raman mapping and electron backscatter diffraction," *Mater. Lett.*, **65**, 19–20, pp. 2864–2867, 2011.

17. T. O. Terasawa and K. Saiki, "Growth of graphene on Cu by plasma enhanced chemical vapor deposition," *Carbon N. Y.*, **50**, 3, pp. 869–874, 2012.

18. U. Narula, C. M. Tan, and C. S. Lai, "Growth mechanism for low temperature PVD graphene synthesis on copper using amorphous carbon," *Sci. Rep.*, **7**, pp. 1–13, 2017.

19. R. W. Hoffman, "Mechanical properties of non-metallic thin films," in *Physics of Nonmetallic Thin Films*, Springer, Boston, MA, 1976.

20. G. G. Stoney, "The tension of metallic films deposited by electrolysis," *Proc. R. Soc. London.*, **82**, 553, pp. 172–175, 1909.

21. S. Xu et al., "Properties of carbon ion deposited tetrahedral amorphous carbon films as a function of ion energy," *J. Appl. Phys.*, **79**, 9, pp. 7234–7240, 1996.

22. I. Ichim, D. V. Kuzmanovic, and R. M. Love, "A finite element analysis of ferrule design on restoration resistance and distribution of stress within a root," *Int. Endod. J.*, **39**, 6, pp. 443–452, 2006.

23. I. Blech and U. Cohen, "Effects of humidity on stress in thin silicon dioxide films," *J. Appl. Phys.*, **53**, 6, pp. 4202–4207, 1982.

24. F. C. Marques, R. G. Lacerda, A. Champi, V. Stolojan, D. C. Cox, and S. R. P. Silva, "Thermal expansion coefficient of hydrogenated amorphous carbon," *Appl. Phys. Lett.*, **83**, 15, pp. 3099–3101, 2003.

25. G. K. White, "Thermal expansion of reference materials: copper, silica and silicon," *J. Phys. D. Appl. Phys.*, **6**, 17, pp. 2070–2078, 2002.
26. D. Yoon, Y.-W. Son, and H. Cheong, "Negative thermal expansion coefficient of graphene measured by raman spectroscopy," *Nano Lett.*, **11**, 8, pp. 3227–3231, 2011.
27. M. T. Kim, "Influence of substrates on the elastic reaction of films for the microindentation tests," *Thin Solid Films*, **283**, 1–2, pp. 12–16, 1996.
28. S. Cho, I. Chasiotis, T. A. Friedmann, and J. P. Sullivan, "Young's modulus, Poisson's ratio and failure properties of tetrahedral amorphous diamond-like carbon for MEMS devices," *J. Micromech. Microeng.*, **15**, 4, pp. 728–735, 2005.
29. M. O. Bloomfield, D. N. Bentz, and T. S. Cale, "Stress-induced grain boundary migration in polycrystalline copper," *J. Electron. Mater.*, **37**, 3, pp. 249–263, 2008.
30. R. Faccio, P. A. Denis, H. Pardo, C. Goyenola, and Á. W. Mombrú, "Mechanical properties of graphene nanoribbons," *J. Phys. Condens. Matter*, **21**, 28, p. 285304, 2009.
31. R. de Groot, M. C. R. B. Peters, Y. M. de Haan, G. J. Dop, and A. J. M. Plasschaert, "Failure stress criteria for composite resin," *J. Dent. Res.*, **66**, 12, pp. 1748–1752, 1987.
32. M. J. Buehler, A. Hartmaier, and H. Gao, "Constrained grain boundary diffusion in thin copper films," *MRS Proceedings*, **821**, p1.2, 2004.
33. J. Yao and J. R. Cahoon, "Experimental studies of grain boundary diffusion of hydrogen in metals," *Acta Metall. Mater.*, **39**, 1, pp. 119–126, 1991.
34. I. Vlassiouk et al., "Role of hydrogen in chemical vapor deposition growth of large single-crystal graphene," *ACS Nano*, **5**, 7, pp. 6069–6076, 2011.
35. H. Ji et al., "Graphene growth using a solid carbon feedstock and hydrogen," *ACS Nano*, **5**, 9, pp. 7656–7661, 2011.
36. Y. Zhang, Z. Li, P. Kim, L. Zhang, and C. Zhou, "Anisotropic hydrogen etching of chemical vapor deposited graphene," *ACS Nano*, **6**, 1, pp. 126–132, 2012.
37. S. Choubak, M. Biron, P. L. Levesque, R. Martel, and P. Desjardins, "No graphene etching in purified hydrogen," *J. Phys. Chem. Lett.*, **4**, 7, pp. 1100–1103, 2013.
38. N. Saito, N. Murata, K. Tamakawa, K. Suzuki, and H. Miura, "Stress-induced migration of electroplated copper thin films used for 3D

integration," in *2010 5th International Microsystems Packaging Assembly and Circuits Technology Conference*, 2010, pp. 1–4.

39. Y. Hou and C. M. Tan, "Comparison of stress-induced voiding phenomena in copper line-via structures with different dielectric materials," *Semicond. Sci. Technol.*, **24**, 8, p. 085014, 2009.

40. Y. Hou and C. M. Tan, "Stress-induced voiding study in integrated circuit interconnects," *Semicond. Sci. Technol.*, **23**, 7, p. 075023, 2008.

41. J. E. Lee, G. Ahn, J. Shim, Y. S. Lee, and S. Ryu, "Optical separation of mechanical strain from charge doping in graphene," *Nat. Commun.*, **3**, 1, pp. 1–8, 2012.

42. F. Ding et al., "Stretchable graphene: A close look at fundamental parameters through biaxial straining," *Nano Lett.*, **10**, 9, pp. 3453–3458, 2010.

43. E. M. Zielinski, R. P. Vinci, and J. C. Bravman, "Effects of barrier layer and annealing on abnormal grain growth in copper thin films," *J. Appl. Phys.*, **76**, 8, pp. 4516–4523, 1994.

44. M. Moriyama, K. Matsunaga, and M. Murakami, "The effect of strain on abnormal grain growth in Cu thin films," *J. Electron. Mater.*, **32**, 4, pp. 261–267, 2003.

Chapter 4

Statistical Approach to Identify Key Growth Parameters of the Novel Graphene Growth PVD Processes

With the mechanism of physical vapor deposition (PVD)-based graphene growth identified as discussed in the previous chapter, it is also necessary to understand the key factors that influence the growth process so that the PVD process can be engineered and one can obtain graphene with different qualities suitable for different applications. The design of experiment (DoE) method, which is based on the statistical approach, is an important method that helps to understand the key factors and their optimization to obtain the intended process parameters. This chapter will provide a brief history of DoE with its need in the current world, which will be further followed by its importance and various applications. Qualitative and quantitative DoE analysis to identify key parameters for this PVD-based graphene synthesis will be discussed in detail. This chapter will also explain a novel filtered quantitative DoE analysis derived from quantitative analysis, which is an addition to the existing techniques.

Graphene and VLSI Interconnects
Cher-Ming Tan, Udit Narula, and Vivek Sangwan
Copyright © 2022 Jenny Stanford Publishing Pte. Ltd.
ISBN 978-981-4877-82-4 (Hardcover), 978-1-003-22488-4 (eBook)
www.jennystanford.com

4.1 Brief History and Need of DoE

Experimental design or design of an experiment or a process, in its literal sense, has been in practice in the society since the very existence of *Homo sapiens*, which dates back to millions of years ago. From small inventions like making fabric for clothes, modifying recipes for food, cooking a perfect cup of tea/coffee, etc. to big revolutionary creations of airplanes, ships, space shuttles, etc., each and every process requires the incorporation of carefully chosen parameters and optimum values to achieve the desired result. However, the terms "Experimental Design" or "Design of Experiment" or "DoE" became a regular exponent of statistical conjecture and deduction for statisticians after Ronald Aylmer Fisher's pioneering and remarkable investigations at the Rothamsted Experimental Station in the 1920s [1]. Since then, this technique is one of the most important tools for statistical inference, focused largely on the agricultural sector during the 1920s to 1940s.

It was only after World War II when the application of DoE saw its reinvigoration, and it found significant popularity in chemical and process development industries. The period between 1950s and 1980s was also marked as the period of the statistical quality revolution, which begun in Japan, and it was administered by Joseph M. Juran [2] and W. Edward Deming [3], a physicist.

Furthermore, the application of DoE has increased in the modern era since the 1990s, especially in industries. This was after Genichi Taguchi, a Japanese engineer, realized that the conventional Fisher's methods were mainly applicable to long-term experiment in order to obtain improved harvests, but needed modification for industrial applications, which were shorter in targets for a defined outcome. Along with many concepts, the most important idea he popularized in the world of statistics was the concept of orthogonal arrays [4] developed by C. R. Rao [5, 6], R. C. Bose and K. A. Bush [7]. Figure 4.1 shows a summary of the brief history of concepts introduced for DoE in industrial applications.

In the synthesis of materials where a small change in the parameters can make a significant impact on the outcomes, the study of material science from the point of view of synthesizing different materials and their growth mechanisms has a large scope of reaping the benefits and advantages that DoE and its concepts offer.

However, the use and application of DoE concepts in this discipline have been very limited despite a wide variety of applications of DoE in industrial processes. This highlights the gap between the physics that governs a process and the underlying statistics of the same process, in the eyes of engineers and scientists working on particular physical processes. In fact, the reality is that the physics of any process goes hand in hand with the associated statistics.

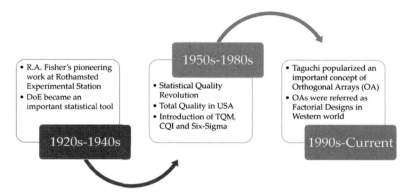

Figure 4.1 Brief history of concepts introduced for DoE in industrial applications.

Out of limited reports on the existence of DoE technique in material science, some of them are process parameters control for a variety of nano-suspensions [8] and nanocarriers [9], surface roughness prediction [10], graphene synthesis by chemical vapor deposition [11], etc.

4.2 Importance of DoE

An experiment is a sequence of tests that are performed in a particular order to enhance or understand the fundamentals of the prevailing process or to discover more about the process or product. DoE is an instrument that can help to understand, enhance, and develop our knowledge about experimentation strategy in order to maximize the quality and quantity of the output using optimum input resources.

Contemporary modern technologies use precious and expensive resources such as rare earth materials or chemicals to develop new products, and at the same time, in the harsh market competition, everyone is competing with each other to beat the clock while maintaining the reasonable price of the product. Therefore, they need to develop experiment processes that can produce high-quality products at less cost of raw materials and in a shorter span of time. In this juncture, DoE plays a crucial role in the production line to determine the crucial factors in an efficient manner, as now we cannot wait to investigate the numerous factors that affect the multifarious processes using a trial-and-error method. In real world, all factors interact with each other and their interactions should be taken in account while designing the experiments. Although conventional techniques of one-factor-at-a-time experiments are easy to perform and understand the effect of individual factors, it is not sufficient to explain the effect of the interactions of factors in the existence of many other factors, which is more important in most of the cases.

DoE can play a vital role in the designing of new experiments or processes where the interaction of multiple factors is not known due to the lack of available scientific information, theory, or principles. It helps to produce new processes in a confident manner and a cost-effective way.

DoE can also be implemented in programs such as the design for reliability (DfR) programs that allow instantaneous analysis of the effect of multiple factors and enable the optimization of the design process.

4.3   Applications of DoE

The DoE technique is implemented in various versatile applications, and in general, it is in more demand nowadays as compared to the time of its origin. It happens due to various reasons such as availability of user-friendly statistical simulation software, which makes the tiring process of equation solving at ease; necessity to produce new processes or recipes to obtain new products in a short time; enhancing the quality of product/process; and optimization of products/processes.

Research on DoE has been done all over the globe from the time it was born, i.e., 1920 to 2019, as shown in Fig. 4.2. An exponential plot can be observed, which displays its importance in the contemporary research society. In the 1960s and 1970s, researchers noticed the significance of DoE to obtain the optimal results in the industries. However, due to the absence of trained people and the unavailability of commercial software, its use was limited, and it was overcome in the 1990s with the advancement of the software industries. Figure 4.3 shows the DoE applications in different scientific communities. The areas benefited the most are medicine and biochemistry, genetics, and molecular biology, which took the most leverage of this technique.

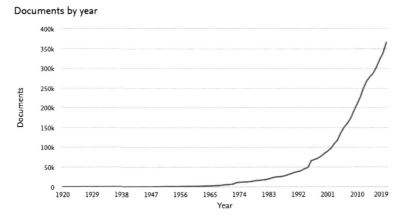

Figure 4.2 Research published in the field of DoE.

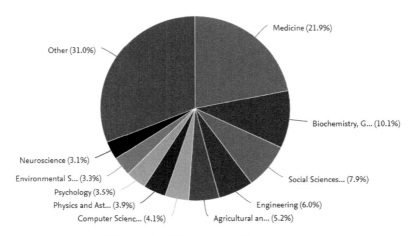

Figure 4.3 Use of DoE by the different scientific community.

4.4 Illustration of DoE for the Novel PVD Graphene Synthesis

4.4.1 Attribute-Response Factorial Design

Attribute-response [12] factorial designs are basically meant to perform qualitative DoE [13]. This type of design looks into the key parameters that collectively influence the possibility of the graphene growth process using the PVD-based method [14].

Attribute-response DoE considers only a proportion of test samples as the output response, on which graphene is found to be synthesized. Hence, there is a need to carefully choose the factors to be considered for the attribute-response analysis. The important thing to note here is that only those factors should be chosen whose variation will not lead to cases where the graphene growth does not take place. This is because such cases will result in null values or "0%," which are not acceptable for attribute-response analysis. The remaining parameters are of constant value as they can keep the graphene growth in a window of possibility. For illustration, three factors/parameters used during post-PVD annealing are considered. These parameters are the a-C layer thickness of sample and temperature–time during the post-PVD annealing process, as depicted in Table 4.1, with their respective levels and labels used for the analysis.

Table 4.1 DoE factors for qualitative and qualitative analysis

	Factors		
Levels	Annealing Temperature (A)	a-C Layer Thickness (B)	Annealing Time (C)
Low (−1)	820°C	12 nm	10 min
High (+1)	1020°C	36 nm	50 min

Source: Reprinted with the permission from Ref. [13], Copyright 2017, IEEE.

Qualitative DoE analysis is implemented on the design based on attribute-response and full-factorial experiments. Since there are three factors, the total experimental rounds to be conducted are 2^3 (in random order) and a complete set of eight combinations of the factors are chosen, as shown in Table 4.2.

Table 4.2 Qualitative and quantitative full-factorial DoE analysis

	Controlling Factors				Peaks in Raman Spectra				Output Responses		
									Qualitative DoE	Quantitative DoE	
	Annealing Temperature	a-C Layer Thickness	Annealing Time	C→AB	I_D	I_G	I_{2D}	(p)g	arcsin(pg)	I_D/I_G peak intensity ratio	I_{2D}/I_G peak intensity ratio
Runs	(A)	(B)							(R1)	(R2)	(R3)
5	820°C	12 nm	10	√	√	√	√	1	1.57	0.600	0.300
7	1020°C	12 nm	10	√	√	√	√	1	1.57	0.400	1.900
1	820°C	36 nm	10	×	×	×	×	0	0	0.001	0.001
6	1020°C	36 nm	10	√	√	√	√	1	1.57	1.000	0.001
3	820°C	12 nm	50	√	√	√	√	1	1.57	0.434	0.434
8	1020°C	12 nm	50	√	√	√	√	1	1.57	0.100	0.250
4	820°C	36 nm	50	√	√	√	×	0	0	0.300	0.001
2	1020°C	36 nm	50	√	√	√	√	1	1.57	0.464	0.890

Source: Reprinted with the permission from Ref. [13], Copyright 2017, IEEE.

Minitab 17 software is employed for DoE analysis of the data recorded in Table 4.2. The I_D, I_G, and I_{2D} peaks, as shown in Table 4.2, are obtained from Raman spectra as they form the three signature peaks in order to identify the presence of graphene [15]. Generally, the peak intensity ratios I_{2D}/I_G and I_D/I_G are defined as metrics for evaluating the number of layers and quality of graphene, respectively [15]. Therefore, these metrics are marked with "X" or "√" depending on the absence and presence of graphene. The term "pg" in Table 4.2 is the proportion of spots with three signature peaks present in the Raman spectrum. The absence of graphene is represented by a "0" value for pg, whereas the presence of graphene on all the spots chosen for analysis is marked by pg equal to "1." In order to normalize the values of pg for DoE analysis, it needs to be converted to nonzero figures. This is done by taking square roots of the arcsine (or inverse sine: $\sin^{-1}$) transform. The arcsin (pg) is the main output response (R1) here, which can be used for further analysis.

The process of DoE analysis begins with the understanding of the main effect plots and interaction plots. The main effect plots help in studying the average change in a particular response as a result of a change in the level of a factor [17]. Nonparallel and steep nature of the main effect plots represents the degree of main effects [17, 18]. On the other hand, interaction plots indicate the disproportionality in the variation in responses across different levels of a factor at given levels of other factors [17]. Nonparallel and crossing plots are an indication of significant interaction [17, 19, 20].

Figures 4.4 and 4.5 show the main effect plots and interaction plots for the attribute-response DoE analysis of R1.

The nonparallel nature of factors A and B in Fig. 4.4 shows that they have a significant effect on the output response R1, while factor C, being parallel to the x-axis, depicts its non-participation in affecting R1. In other words, the a-C layer thickness in the sample and temperature during the post-PVD annealing process have major impact on the response in comparison to annealing time. As it is desirable to have a large value of R1, the main effect plot indicates a high value of temperature during annealing and thin a-C will have more chances of successful graphene growth using the said method.

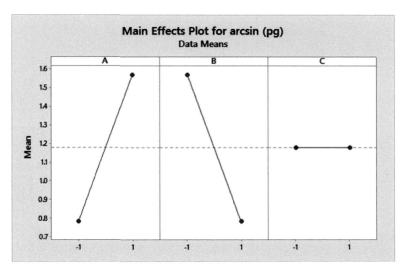

Figure 4.4 Main effect plots for attribute-response DoE analysis of R1. Reprinted with the permission from Ref. [13], Copyright 2017, IEEE.

Furthermore, a clear and substantial interaction can be observed between factors A and B in Fig. 4.5, indicating that the temperature of the chamber during post-PVD annealing has significant interaction with the a-C layer thickness in the sample. In other words, when the a-C layer thickness (factor B) is decreased from a higher level to a lower level, response R1 will remain constant if factor A or the annealing temperature is high. However, response R1 will rise in the same scenario if factor A is low. Hence, the highest value of factor A suggested by the main effect plot would not be optimum for the highest value of response R1 and better success rate for graphene growth. Therefore, response optimization is carried out to obtain the optimum value for the maximized response R1, using the best fit for the response residual equation.

Figure 4.6 shows the results of the response optimizer built-in tool in Minitab v17. In Fig. 4.6, factors A, B, and C are indicated to have values 0, −1, and 1, respectively. Hence, the optimized values for chamber temperature during post-PVD annealing are midway between the high/low levels, which would be 920°C; the thickness of the a-C layer is 12 nm, and the annealing time is 50 min. These optimized parameters are projected to predict an output response of the synthesis of graphene with 95% confidence level.

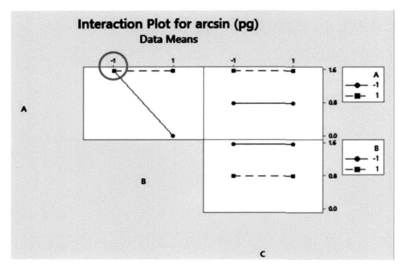

Figure 4.5 Interaction plots for attribute-response DoE analysis of R1.

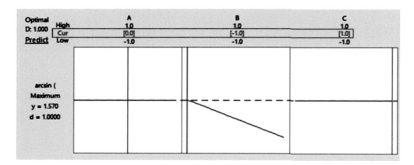

Figure 4.6 Optimum parameters for the maximized response R1.

In order to verify the optimum parameters predicted by DoE analysis, an experimental run is carried out with the optimized recipe, and the Raman characteristics are shown in Fig. 4.7.

Figure 4.7 shows the power of DoE, which can successfully predict the factors of experiments without even performing them. Intuitively, the optimized recipe is expected to produce graphene with a weak Raman signature as the lower temperature increases the chances of depositing graphitic or sp^3 carbon as well. Contrary to this, the Raman signature comes out to be very prominent, which opens up ways to perform graphene synthesis using the said method at lower temperatures as well.

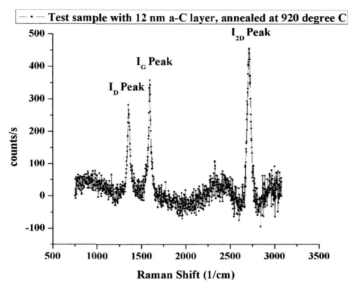

Figure 4.7 Raman characteristics for an experimental run with optimized parameters.

4.4.2 Full-Factorial DoE Analysis

Qualitative DoE analysis presented in the previous section is really handy in determining the best case-optimum factors, which will hold a better chance to synthesize graphene in given conditions. Not only one can obtain this key information; it can also open new avenues for the growth dynamics. However, qualitative DoE analysis lacks the element of analyzing key factors that will influence the quantitative aspects of graphene growth, which are primarily the defect density and the number of synthesized graphene layers. This is quite obvious since the attribute-response DoE analysis considers only the proportion of spots on a given sample having or not having graphene grown on them, and hence the information about the two aforementioned quantitative aspects is missing. Therefore, there is a need for a quantitative DoE analysis for graphene growth.

The methodology of quantitative DoE analysis is quite similar to the attribute-response DoE. The number of factors, their labels, and levels are similar to qualitative analysis, as shown in Tables 4.1 and 4.2. Some exceptions for the quantitative DoE analysis are as follows:

- The quantitative DoE analysis again employs full-factorial DoE but with two output responses R2 and R3, which are named "I_D/I_G peak intensity ratio" and "I_{2D}/I_G peak intensity ratio," respectively.
- The total number of experimental runs performed can be calculated as:

  $n \times L^F = 2 \times 2^3 = 16$, in a random order, where

  i. n is the number of replicates,
  ii. L is the number of levels, and
  iii. F is the number of factors.

- Responses R2 and R3 obtained from Raman characterization for each of the eight sets of factor levels (each having two replicates) are collected, averaged, and recorded, as shown in Table 4.2.

Similar to qualitative DoE analysis, the main effect and interaction plots are studied here.

Figures 4.8 and 4.9 show the main effect and interaction plots of all the three factors for the two responses R2 and R3.

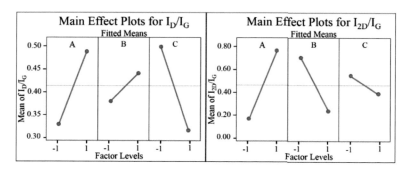

Figure 4.8 Main effect plots for quantitative DoE analysis of R2 (left) and R3 (right). Reprinted with the permission from Ref. [13], Copyright 2017, IEEE.

Illustration of DoE for the Novel PVD Graphene Synthesis | 77

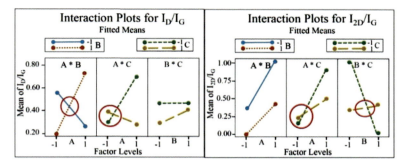

Figure 4.9 Interaction plots for quantitative DoE analysis of R2 (left) and R3 (right). Reprinted with the permission from Ref. [13], Copyright 2017, IEEE.

In Fig. 4.8, the main effect plots for all the three factors seem to affect both the responses R2 and R3. The interaction plots in Fig. 4.9 suggest that there are significant interaction terms, namely, AB and AC for response R2, and AC and BC for response R3.

The individual and interactive changes observed in the two responses as a result of the variation in factor level could arise due to experimental noise. Likewise, significant interaction terms are known to mask out the significant main effects [21]. Hence, only statistically significant changes can be considered main or interaction effects, which need to be analyzed with the DoE plots concurrently.

The statistical significance can be tested using analysis of variance (ANOVA) statistics. Tables 4.3 and 4.4 show the ANOVA analysis for the two responses R2 and R3, respectively.

The statistically significant term should be less than 0.1, which is represented by the "p" value in the ANOVA analysis, which corresponds to 10% significance level. In the case of response R2, the factors' interaction "AB" is the only statistically significant term because its "p" value is 0.062 (less than 0.1), as shown in Table 4.3. On the other hand, the main effects of the factors "A," "B," "C," and the factors' interaction terms "AC," "BC" are all statistically insignificant because their "p" values are more than 0.1, as depicted from the plots in Figs. 4.8 and 4.9.

Table 4.3 ANOVA statistics for response R2 [16]

Factors	Degree of Freedom (DF)	Adjusted Sum of Squares (SS)	Adjusted Mean of Squares (MS = SS/DF)	F Statistics = MS/MSE	P Value	Contribution
Linear Effects							
A	1	0.0495	0.0495	0.65	0.445	7.46%
B	1	0.0067	0.0067	0.09	0.775	1.01%
C	1	0.0618	0.0618	0.80	0.396	9.31%
Two-Way Interactions							
AB	1	0.3599	0.3599	4.69	0.062	54.28%
AC	1	0.1174	0.1174	1.53	0.251	17.70%
BC	1	0.0066	0.0066	0.086	0.776	0.99%
Error	8	0.6142	MSE = 0.076			9.26%

Note: The statistically significant terms are marked in red color.

Table 4.4 ANOVA statistics for response R3 [16]

Factors	Degree of Freedom (DF)	Adjusted Sum of Squares (SS)	Adjusted Mean of Squares (MS = SS/DF)	F Statistics = MS/MSE	P Value	Contribution
			Linear Effects			
A	1	0.6641	0.6641	5.95	0.040	22.44%
B	1	0.4955	0.4955	4.44	0.068	16.74%
C	1	0.0491	0.0491	0.44	0.525	1.66%
			Two-Way Interactions			
AB	1	0.0347	0.0347	0.31	0.592	1.17%
AC	1	0.1001	0.1001	0.90	0.371	3.38%
BC	1	0.7230	0.7230	6.48	0.034	24.43%
Error	8	0.8931	MSE = 0.111			30.18%

Note: The statistically significant terms are marked in red color.

From Table 4.4, it can be found that factors "A," "B," and the interaction term "BC" are all statistically significant for response R3 at the 10% significance level, as their "p" values are 0.040, 0.068, and 0.034, respectively. Among them, the interaction term "BC" is the most significant term followed by the factors "A" and "B," as depicted from the plots in Figs. 4.8 and 4.9.

Table 4.5 Map of choice of levels and quality and number of graphene layers [16]

Desired Graphene Traits	Responses R2 and R3	Factors A Annealing Temperature	B a-C Layer Thickness	C Annealing Time
Low defects only	Low R2	Low for high B; High for low B	Low	Large
Mono-/few-layer graphene only	High R3	High	Low	Small
Mono-/few-layer graphene with low defects	Low R2 and high R3	High	Low	Need to be optimized using response surface method [21, 23]
Multi-/many-layer graphene only	Low R3	Large region in Fig. 4.10, representing several combinations		
Multi-/many-layer graphene with low defects	Low R2 and low R3	High	Low	Large

Illustration of DoE for the Novel PVD Graphene Synthesis | 81

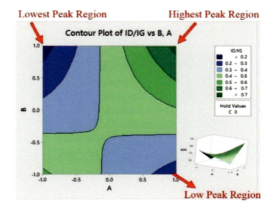

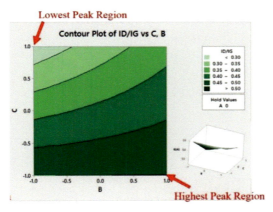

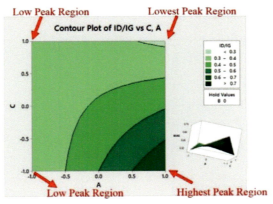

Figure 4.10 Contour plot analysis for response R2 [16].

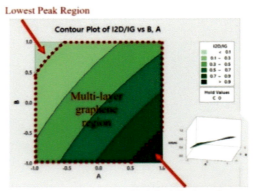
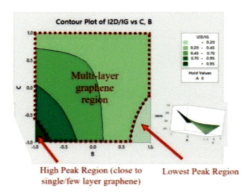
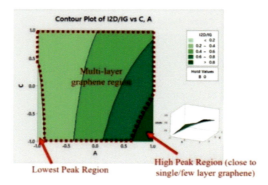

Figure 4.11 Contour plot analysis for response R3 [16].

Apart from knowing the significant factors and their interactions, the optimal parameters for good-quality graphene growth, i.e., low I_D/I_G peak intensity ratio (R2), and the different number of graphene layers, i.e., variable I_{2D}/I_G peak intensity ratio (R3), can be determined. The contour plot analysis based on the results from DoE is drawn, and they are shown in Figs. 4.10 and 4.11 for responses R2 and R3, respectively [16].

By understanding the trends in Figs. 4.10 and 4.11, one can draw out a clear map of different choice of levels for the factors corresponding to different quality and number of graphene layers, as shown in Table 4.5.

In order to further validate the values suggested in Table 4.5, the time for annealing is varied from 5 to 50 min, keeping the samples with a-C layer thickness of 12 nm (low) and post-PVD annealing temperature of 1020°C (high), and the responses are studied using Raman characteristics data. The results are summarized in the plots of I_D/I_G and I_{2D}/I_G peak intensity ratios versus annealing time in Fig. 4.12.

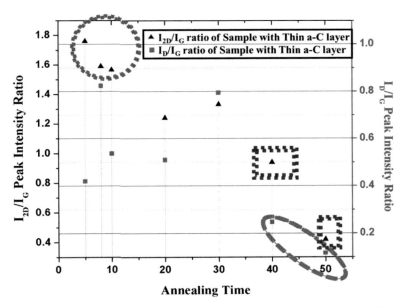

Figure 4.12 Plots of I_D/I_G and I_{2D}/I_G peak intensity ratios versus annealing time, for samples with thin a-C layer annealed at 1020°C [16].

The region of larger I_{2D}/I_G peak intensity ratio (an indication of many/multilayer graphene) at shorter annealing times represented by a dotted circle, the region with smaller I_{2D}/I_G peak intensity ratio (an indication of single-/few-layer graphene) shown by a dotted rectangle, and the region with smaller I_D/I_G peak intensity ratio (an indication of graphene with low defect density) shown by dashed oval in Fig. 4.12 are in agreement with the optimal recipe map (in Table 4.5) suggested by DoE.

4.4.3 Filtered DoE Approach

From the qualitative and quantitative DoE analysis results, it can be concluded that the thickness of a-C layer in the experimental sample and the temperature during post-PVD annealing are significant for both the feasibility of graphene growth and its quality. However, the time factor for the post-PVD annealing process is only significant for the quantitative DoE analysis. This indicates that the quality of graphene is largely dependent on the variation in the annealing time.

Hence one can filter the factor levels according to previous DoE analysis results to obtain finer detailed information. In other words, one can fix one of the parameters, which is not so significant, and perform DoE on a reduced number of parameters. If it is hard to decide which parameter to fix, response optimizer tools can be used to start with. This technique is named filtered quantitative DoE analysis. Figure 4.13 shows the optimized recipe by the response optimizer tool of Minitab where response R2 is forced to be minimum and response R3 is forced to be maximum in order to study parameters that can produce single-/few-layer graphene.

These suggested values are shown in Fig. 4.13, which translates into temperature and time during post-PVD annealing of about 920°C and 20 min, respectively, and a thickness of a-C layer of 36 nm. This suggested value is, at first, quite surprising as for single-/few-layer graphene, a thin a-C layer is preferred, as shown in Table 4.5. But on careful observation, it can be seen that the suggested values of the other factor levels are totally different from the ones used in Table 4.5. This again shows the power of DoE, that it can suggest a pool of many combinations fit for a targeted process.

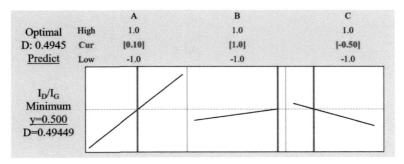

Figure 4.13 Optimum recipe for minimum I_D/I_G peak and maximum I_{2D}/I_G peak intensity ratios. Reprinted with the permission from Ref. [13], Copyright 2017, IEEE.

Considering the filtering approach, the best choice for keeping one of the factor constants comes from the ease and feasibility. The thickness of the a-C layer is the most time-consuming factor out of all the three parameters. This is because the temperature and time during the annealing process can be tuned easily, whereas tuning the a-C layer thickness requires changing of several parameters in the RF sputtering system. Consequently, the preparation of samples with a given a-C layer thickness followed by tuned parameters of time and temperature during annealing is simpler than samples with different a-C layer thickness. Hence, in order to reduce the number of factors, a-C layer thickness is kept constant to 36 nm as suggested by the response optimizer, and DoE is carried out using annealing temperature and time, as shown in Tables 4.6 and 4.7.

From the filtered DoE approach, one can create recipes capable of synthesizing good-quality and defect-free graphene.

Table 4.6 Filtered DoE factors

	Factors for a-C Thickness, B = 36 nm	
	Annealing Temperature	**Annealing Time**
Levels	Ann. Temp	Ann. Time
Low (−1)	920°C	20 min
High (+1)	1020°C	50 min

Source: Reprinted with the permission from Ref. [13], Copyright 2017, IEEE.

Table 4.7 Output responses for filtered DoE analysis

	Sample with 36 nm a-C Thickness			
Graphene Signature Peak Intensity Ratio	Annealed at 1020°C		Annealed at 920°C	
	For 50 min Run order: 4	For 20 min Run order: 2	For 50 min Run order: 3	For 20 min Run order: 1
I_D/I_G response R2A	0.5556 high defect	0.1786 low defect	0.5534 high defect	0 no or negligible defect
I_{2D}/I_G response R3A	0.9000 multilayer graphene	0.9143 multilayer graphene	1.515 few-layer graphene	2.6 single-layer graphene

Source: Reprinted with the permission from Ref. [13], Copyright 2017, IEEE.

4.5 Summary

DoE analysis performed in graphene synthesis is an interesting approach as there are many dimensions to the material growth that one can explore. The summary of the DoE application is represented by a schematic in Fig. 4.14.

The downward conical shape of the summarized schematic in Fig. 4.14 signifies the capability and benefits of DoE in terms of traversal from qualitative analysis, which is wrapped around uncertainties to a more precise quantitative approach. Furthermore, it leads to filtered quantitative analysis, which is highly directional in terms of identifying key parameters for a successful and effective process as intended.

The qualitative DoE analysis is useful in determining factors affecting the feasibility of graphene growth, and the thickness of a-C layer in the sample and temperature during the post-PVD annealing process are found to be the critical ones in this regard for the PVD-based graphene synthesis method. This allows us to proceed further to quantitative DoE analysis for determining the factors affecting the quality of synthesized graphene in terms of the number of graphene layers and the defect density of the crystal, revealing different routes or recipes for the variable quality of graphene.

Summary

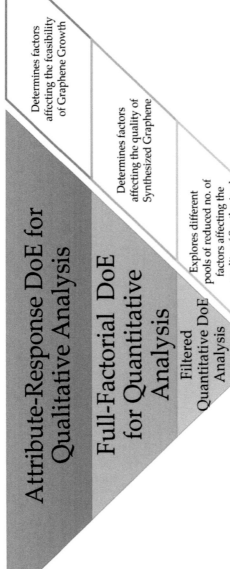

Figure 4.14 Summary of application of DoE in the graphene synthesis process.

In order to explore further into the optimal growth parameters while keeping the number of experiments reasonable, the filtered quantitative DoE analysis is developed. This new way of analysis uses a derivative of different levels of parameters suggested by full-factorial quantitative DoE analysis and obtains multiple recipes for a particular quality of graphene. An optimal recipe of moderate annealing temperature, thick a-C layer, and faster annealing time in the case of PVD-based graphene synthesis method is found to produce graphene with negligible defect density in its Raman signature.

References

1. E. J. Russell, "'Sir Ronald Fisher; Mathematical Biology', Letter to The Times of London," [Online], 1962.
2. J. M. Juran, "Japanese and western quality: A contrast," *Qual. Prog.*, **11**, 12, pp. 10–18, 1978.
3. W. E. Deming, "What happened in Japan?" *Ind. Qual. Control*, **24**, 2, 1967.
4. G. Taguchi and S. (Shōzō) Konishi, *Orthogonal Arrays and Linear Graphs: Tools for Quality Engineering*, 1987.
5. C. Rao, "On hypercubes of strength d and a system of confounding in factorial experiments," *Bull. Calcutta Math. Soc*, **38**, pp. 67–78, 1946.
6. C. Rao, "Factorial experiments derivable from combinatorial arrangements of arrays," *Suppl. J. R. Stat. Soc*, **9**, 1, p. 128, 1947.
7. R. C. Bose and K. A. Bush, "Orthogonal arrays of strength two and three," *Ann. Math. Stat.*, **23**, pp. 508–524, 1952.
8. S. Verma, Y. Lan, R. Gokhale, and D. J. Burgess, "Quality by design approach to understand the process of nanosuspension preparation," *Int. J. Pharm.*, **377**, 1–2, pp. 185–198, 2009.
9. X. Xu, M. A. Khan, and D. J. Burgess, "A quality by design (QbD) case study on liposomes containing hydrophilic API: I. Formulation, processing design and risk assessment," *Int. J. Pharm.*, **419**, 1–2, pp. 52–59, 2011.
10. I. A. Choudhury and M. A. El-Baradie, "Surface roughness prediction in the turning of high-strength steel by factorial design of experiments," *J. Mater. Process. Technol.*, **67**, 1–3, pp. 55–61, 1997.

11. C. Wirtz, K. Lee, T. Hallam, and G. S. Duesberg, "Growth optimisation of high quality graphene from ethene at low temperatures," *Chem. Phys. Lett.*, **595–596**, pp. 192–196, 2014.

12. M. Jayakumar, "Design of experiments (DoE)," [Online]. Available: https://www.isixsigma.com/tools-templates/design-of-experiments-doe/. [Accessed: 05-Jan-2018].

13. U. Narula and C. M. Tan, "Engineering a PVD-based graphene synthesis method," *IEEE Trans. Nanotechnol.*, **16**, 5, pp. 784–789, 2017.

14. U. Narula, C. M. Tan, and C. S. Lai, "Growth mechanism for low temperature PVD graphene synthesis on copper using amorphous carbon," *Sci. Rep.*, **7**, pp. 1–13, 2017.

15. A. C. Ferrari, "Raman spectroscopy of graphene and graphite: Disorder, electron-phonon coupling, doping and nonadiabatic effects," *Solid State Commun.*, **143**, 1–2, pp. 47–57, 2007.

16. U. Narula and C. M. Tan, "Determining the parameters of importance of a graphene synthesis process using design-of-experiments method," *Appl. Sci.*, **6**, 7, pp. 1–16, 2016.

17. J. Spall, "Factorial design for efficient experimentation," *IEEE Control Syst. Mag.*, **30**, 5, pp. 38–53, 2010.

18. G. E. P. Box, J. S. Hunter, and W. G. Hunter, *Statistics for Experimenters: Design, Innovation, and Discovery*. Wiley-Interscience, 2005.

19. B. Thomas and A. Milivojevich, *Quality by Experimental Design*. CRC Press, 2016.

20. D. C. Montgomery and G. C. Runger, *Applied Statistics and Probability for Engineers*. Jefferson, USA: John Wiley & Sons, Inc., 2010.

21. D. C. Montgomery, *Design and Analysis of Experiments*. John Wiley & Sons, Inc., 2013.

22. M. Yolmeh and S. M. Jafari, "Applications of response surface methodology in the food industry processes," *Food Bioprocess Technol.*, **10**, 3, pp. 413–433, 2017.

23. M. Sarfarazi, S. M. Jafari, and G. Rajabzadeh, "Extraction optimization of saffron nutraceuticals through response surface methodology," *Food Anal. Methods*, **8**, 9, pp. 2273–2285, 2015.

Chapter 5

Copper–Graphene Interconnect

5.1 Introduction

Given the excellent electrical and thermal properties of graphene, and the demonstration of the possibility of growing graphene on copper using physical vapor deposition (PVD) process in Chapter 3, the graphenated Cu can be a viable candidate for future very large-scale integration (VLSI) interconnections. This chapter presents the electrical and thermal properties of graphenated Cu through experiments. This chapter also presents the improved electromigration (EM) performance of such graphenated copper interconnects through atomic level finite element modeling of EM.

An interesting property of graphene that allows electroless Cu deposition without the use of any other chemicals or reducing agents has been discovered. Conventionally, electroless metal deposition requires the use of other additives or reducing agents. Avoiding such chemicals not only saves cost but also solves the problem of disposing of such chemicals, which are known to be non-biodegradable and toxic in nature. With such a newly founded property of graphene, a novel metal–graphene–metal interconnect can be possible, and its advantage will be discussed in this chapter.

5.2 Electrical and Thermal Characteristics of Graphenated Copper

5.2.1 Sample Description

Using the PVD method with the same conditions described in Section 2.1.1 of Chapter 2, 400 nm of copper (Cu) is deposited onto SiO_2/Si substrate. Alongside this sample, another sample is prepared similar to sample S1.2 described in Section 2.2.1 in Chapter 2, for comparison. The Cu thickness for this sample is 400 nm, amorphous carbon thickness is 12 nm, and the subsequent annealing is performed at 920°C to form the graphenated Cu sample.

5.2.2 Experimentation and Results

The electrical and thermal properties of the samples are measured using a four-point probe method and thermal scanner, respectively.

5.2.2.1 Temperature distribution measurement

Figure 5.1 shows the thermal scanner images (collected using P384A-20 Series high-performance thermos-image camera from Ching Hsing Computer-Tech Ltd.) of graphenated Cu and Cu thin film on Si/SiO_2 substrate, respectively. These samples are powered by current probes with a value of 3 A. It is clearly seen in Fig. 5.1 that the temperature of graphenated Cu is in the range of 32–35°C, which is lower as compared to that of Cu thin film (36–41°C). Also, the temperature distribution in the graphenated Cu is a lot more uniform in comparison to Cu thin film sample. The hot spots observed on graphenated Cu are due to the force applied by the probes during a four-point measurement, possibly causing film rupture. The results are expected because graphene is an excellent thermal conductor.

Temperature gradient and induced hydrostatics stress gradient are the two major driving forces in the EM reliability of narrow interconnects (less than 100 nm width) [1], and these main factors depend directly upon the uniform temperature distribution in the film. Also, with lower temperature, the atomic diffusion rate decreases, and this will further enhance interconnect EM phenomena.

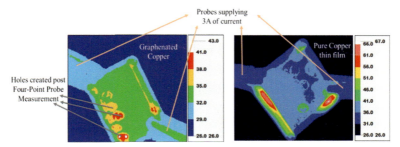

Figure 5.1 Thermal scanner images of graphenated Cu and Cu thin film on Si/SiO$_2$ substrate, respectively. Republished with the permission of John Wiley & Sons from Ref. [2]; permission conveyed through Copyright Clearance Center, Inc.

Interestingly, there is a new finding that the graphenated Cu sample with different graphene quality in terms of defect density has different thermal response. Figures 5.2a,b show the Raman characteristics of graphenated Cu samples prepared by post-PVD annealing (at 920°C) for 20 min and 50 min, respectively. The defect density indicated by the I_D/I_G peak intensity ratio of about 0.66 for 50 min of annealing and almost negligible for 20 min of annealing.

In other words, the samples with some defective domains in graphene are found to have better thermal properties. This can be attributed to the presence of slightly defected graphene domains and multilayered configuration in the grown graphene with 50 min of annealing as observed from its Raman characteristics. These defective domains actually form discontinuous bridges and facilitate graphene-edge connections [4–6] with the Cu surface. Moreover, the multilayered configuration of graphene acts like multiple highways for current and heat flow, enhancing the conduction and thermal paths, respectively.

5.2.2.2 Electrical resistivity measurement

Four-point measurement (in house four-point probe built using sub-parts from 3S Co. Ltd.; measuring unit is by Keithley's 2400 Source-meter) data also suggest that the samples with negligible defect density (suggested from Fig. 5.2) not only have poor thermal properties but also have higher resistivity, as shown in Table 5.1.

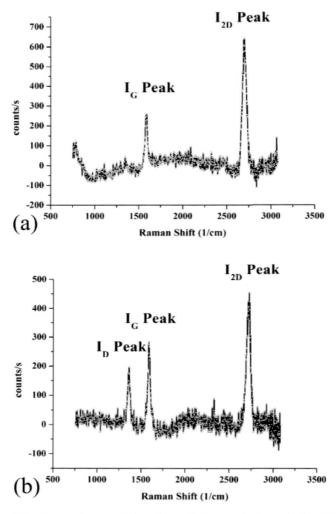

Figure 5.2 Raman characteristics of post-PVD annealed sample for different annealing duration of (a) 20 min and (b) 50 min.

In any case, the conduction performance of graphenated copper is found to be better than pure Cu thin film as shown by resistivity measurements using the four-point system in Table 5.1 where it clearly shows a vast improvement in conductivity values and hence better current-carrying capacity.

Table 5.1 Four-point probe measurement for the experimental sample

Sample	Measured Resistivity (10^{-6} ohm-cm)	Standard Value (10^{-6} ohm-cm)
Graphenated copper obtained by 20 min of annealing	$\rho_{Gr\text{-}Cu\text{-}20\ min} = 1.4391$	—
Graphenated copper obtained by 50 min of annealing	$\rho_{Gr\text{-}Cu\text{-}50\ min} = 1.0935$	—
Cu thin film (300 nm) on Si/SiO$_2$ substrate	$\rho_{Cu} = 3.228548$	1.68–2.2 [3] (only Cu thin film)

Source: Republished with the permission of John Wiley & Sons from Ref. [2]; permission conveyed through Copyright Clearance Center, Inc.

An illustration of high current density tolerance for graphene, in particular, can be given by the following example by using the measurement data in Table 5.1.

Let us suppose an interconnect structure of 800 µm in length (*l*) with 100 nm width (*w*) is constructed with thicknesses of copper (t_{Cu}) and graphene (t_{Gr}) of 400 nm and 0.3 nm, respectively. The resistance values of graphenated copper ($R_{Gr\text{-}Cu}$) and pure copper (R_{Cu}) can be evaluated as shown in Eqs. (5.1) and (5.2):

$$R_{Cu} = \rho_{Cu} \frac{l}{w \times t_{Cu}} = \frac{3.228 \times 10^{-6} \times 800 \times 10^{-4}}{100 \times 10^{-7} \times 400 \times 10^{-7}} = 645.70\ \Omega \quad (5.1)$$

$$R_{Gr\text{-}Cu} = \rho_{Gr\text{-}Cu\text{-}50\ min} \frac{l}{w \times (t_{Cu} + t_{Gr})}$$
$$= \frac{1.4391 \times 10^{-6} \times 800 \times 10^{-4}}{100 \times 10^{-7} \times 400.3 \times 10^{-7}} = 287.82\ \Omega \quad (5.2)$$

Assume 1 V is applied to the interconnect, and according to the previously calculated resistance values of the Cu and graphene film, the current in the graphenated copper ($I_{Gr\text{-}Cu}$) and pure copper (I_{Cu}) will be 3.474 mA and 1.548 mA, respectively. In other words, this graphenated copper has more than double the current-carrying capability.

Moreover, with graphene placed over the copper surface, we can model them as a parallel configuration, as depicted in Fig. 5.3. From the previously calculated resistances and current values in Eqs. (5.1) and (5.2), the current density values can be calculated as:

$$J_{Cu} = I_{Cu}\frac{1}{w \times t_{Cu}} = \frac{1.548 \times 10^{-3}}{100 \times 10^{-7} \times 400 \times 10^{-7}} = 3.87 \times 10^{6} \text{ A/cm}^2 \quad (5.3)$$

$$J_{Gr} = I_{Gr}\frac{1}{w \times t_{Gr}} = \frac{1.926 \times 10^{-3}}{100 \times 10^{-7} \times 0.3 \times 10^{-7}} = 6.42 \times 10^{9} \text{ A/cm}^2 \quad (5.4)$$

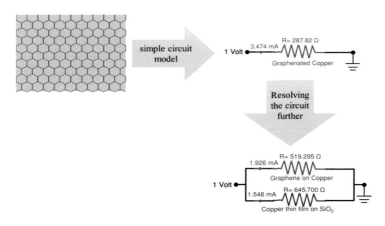

Figure 5.3 Simple circuit model for graphenated copper.

On comparing Eq. (5.3) and Eq. (5.4), it is clear that the current density in graphene is three orders of magnitude higher than that in copper. Therefore, the presence of graphene reduces the current density in copper and further helps in improving its EM lifetime, besides the reduction in temperature and temperature gradient. To examine the impact of such graphene film on Cu on the EM performance, atomic level finite element EM simulation developed by Tan et al. [1, 7] is used for the investigation.

5.2.3 Atomic Level Finite Element EM Modeling

Atomic-based finite element EM modeling in a graphenated interconnect system is performed for M1-M2 test structure using Mechanical APDL (ANSYS Parametric Design Language) v17 tool. The modeling is performed purely using APDL, which is a powerful scripting language that allows to parameterize a model and automate common tasks. This modeling has been adapted from the works done by Tan et al. [1, 7, 8].

Figure 5.4 shows the schematic of the test structure used for the FEM simulation. Tables 5.2 and 5.3 (extracted from Refs. [8–10]) show the model feature sizes and parameters used for modeling.

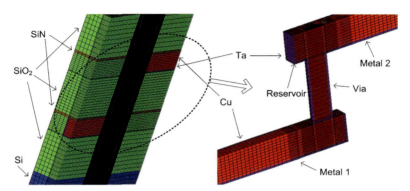

Figure 5.4 Model of the test structure redrawn using work by Tan et al. Reprinted with the permission from Ref. [8], Copyright 2007, AIP Publishing.

Table 5.2 Feature sizes of the test structure used in FEM

Material	Young's Modulus [GPa]	Poisson's Ratio	Thermal Conductivity (W/mK)	Thermal Coefficient of Expansion ×10^{-6}[°C^{-1}]
Cu	129.8	0.339	379	16.0
Ta	186.2	0.35	53.65	6.48
SiN	265	0.27	0.8	1.5
SiO$_2$	71.4	0.16	1.75	0.68
Si	130	0.28	61.9	4.4
Graphene	960	0.17	5000	−8

Source: Reprinted from Ref. [8] with the permission from AIP Publishing.

Abe et al. [11] and Sasagawa et al. [12] provided a detailed explanation of using atomic flux divergence (AFD) calculations to determine the EM. During the void formation, various factors need to be considered; among them AFD is acknowledged as the most important parameter. The M1-M2 test structure model is developed in the ANSYS APDL simulation tool with and without graphene. This test structure is simulated at 300°C high stress-test temperature to calculate the AFD, and the results are shown in Fig. 5.5.

Table 5.3 Model parameters for the test structure

Feature	Size (μm)
Line width	0.28
Line thickness	0.35
Via diameter	0.26
Via height	0.68
Barrier layer thickness	0.025
Cap layer thickness	0.05
Reservoir length	0.125
Silicon substrate thickness	300

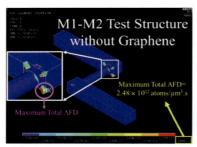
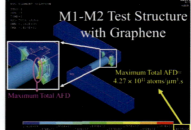

Figure 5.5 Atomic flux divergence calculation for FEM simulation of M1-M2 EM test structure. Republished with the permission of John Wiley & Sons from Ref. [2]; permission conveyed through Copyright Clearance Center, Inc.

It is clearly evident from the simulation results, as presented in Fig. 5.5, that graphene has much better and more promising results in terms of EM performance in comparison to the interconnect structure without graphene. The total AFD of the interconnect with graphene is nearly 1/6th of the one without graphene. Figure 5.6 emphasizes the strength of the graphenated Cu interconnect by showing the resistance change of the interconnect structures after a certain period of time. The test structure without graphene shows a sudden increase in resistance while the same continues to remain low until six more loops.

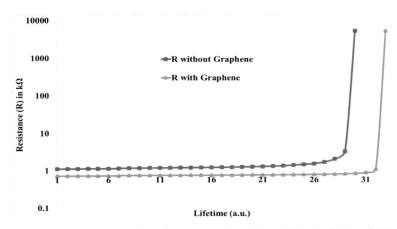

Figure 5.6 Variance in the resistance calculated from FEM simulation of M1-M2 EM test structure with and without graphene.

A well-known proven fact provided by Tan et al. [8] that the average of total averaged AFD is inversely proportional to the lifetime T and given as:

$$T \propto 1/\text{av}(\text{AFD}_{\text{av}_{\text{total}}}) \tag{5.5}$$

A comparison in EM lifetime can be made in determining the lifetime of the test structures by using Eq. (5.5), and the calculation is as follows, where the EM lifetime is improved at least three times with the addition of graphene on Cu.

$$\frac{T_{\text{with Graphene}}}{T_{\text{without Graphene}}} = \frac{\text{av}(\text{AFD}_{\text{av}_{\text{total}}}\text{without Graphene})}{\text{av}(\text{AFD}_{\text{av}_{\text{total}}}\text{with Graphene})} = 3.33 \text{ approximately}$$

$$\tag{5.6}$$

5.3 Compatibility of Graphenated Interconnect to Current Integrated Circuit Back-End Processes

With the promising graphenated copper and considering the standard damascene process for the fabrication of VLSI interconnects, the integration of graphenated Cu interconnects in integrated circuit fabrication, especially with via formation for multi-level interconnections, is a challenge. To overcome the challenge, a practical technique is available that inserts the graphene layer inside copper interconnects as depicted in Fig. 5.7.

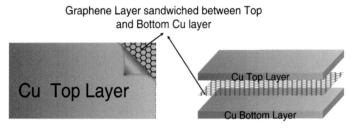

Figure 5.7 Graphene sandwiched within copper interconnection. Republished with the permission of John Wiley & Sons from Ref. [2]; permission conveyed through Copyright Clearance Center, Inc.

There are several methods to put Cu on graphenated copper, namely, electroless Cu deposition [13, 14], electroplating [15], etc. The most common way is to electroplate Cu in interconnect fabrication as a part of the standard damascene process. On the other hand, electroless copper (Cu) deposition is a well-established process that promises huge cost savings and simple process step.

The electroless deposition has remained the mainstay in chemical processes since the 1840s [16] because it holds a considerable significance mainly in the metal deposition industry. In particular, this electroless deposition of the copper method came into picture in the 1970s [14, 17], which employed various types of additives, reducing agents, and chemicals [18–23].

The basic principle behind the electroless process is the electrochemical potential provided by the chemical reaction caused by reducing agents/chemicals/additives to reduce copper. Unfortunately, these additives are generally not environmentally friendly.

Using the same principle, graphenated Cu with suitable ambient conditions can provide sufficient electrochemical energy for the deposition to persist. A novel electroless Cu deposition process is developed for the graphenated Cu.

5.4 Novel Copper–Graphene–Copper Interconnect and Its Potential Performance

Monolayer CVD-graphene on Cu foil (graphenated Cu) was kept in 30 mL copper sulfate (0.2 M $CuSO_4 \cdot 5H_2O$; pH ~ 4) solution for the time durations of 3, 6, 9, and 12 h.

An important and noticeable interaction was observed between the copper sulfate ($CuSO_4$) electrolyte solution and graphenated Cu when graphene sitting on the copper foil is immersed in the electrolyte solution.

Raman characteristics on the samples (i.e., before and after electrolyte treatment) indicate that there is no significant shift in the peak positions, and there is also an absence of I_D peak even after electrolyte interaction [2]. However, lowering of the peak intensities can be clearly seen in the plots, which could be due to the change in the surface morphology of the samples after the electrolyte treatment.

SEM examination on (taken from JEOL-JSM SEM – JSM6700F with OXFORD instruments EDS) samples treated with electrolyte for 3 h gives a good indication of crystal-like island deposition on the surface, as compared to the fresh samples. In the case of a longer interaction period, the coverage of island growth is more [2].

EDS analysis results on the sample surface with its quantitative analysis results shown in Table 5.4 confirm that these crystal-like islands are copper (I) oxide (Cu_2O) crystals [2]. The formation of Cu_2O crystals occurred due to the two main possibilities:

1. Reduction of Cu^{2+} ions to Cu^+ ions in the presence of the electrolyte solution [24].
2. Reduction of Cu^{2+} to Cu inside electrolyte solution followed by native oxidation [25] of deposited copper in the air after taking out the sample.

An aqueous environment was used to perform the experiment; hence, the first possibility is not stable [26, 27]. Any formation of Cu^+ ions will not be possible as these ions will further be reduced to Cu and become oxidized after getting exposed to air. Hence, the second possibility is most likely in this case. Since oxidation of the deposited Cu is not acceptable for integrated circuit fabrication, proper care must be taken by avoiding the contact of oxygen to the freshly formed Cu on the graphenated Cu.

Table 5.4 Average Cu/O atomic % for different samples from EDS

Sample	Region	Average Cu/O Atomic %
Fresh graphenated Cu sample	Graphene on Cu	28.0±7.8
Sample dipped in electrolyte for 3 h	Graphene on Cu	24.9±3.3
	Deposited Cu islands	2.2±0.3
Sample dipped in electrolyte for 12 h	Graphene on Cu	25.6±6.1
	Deposited Cu islands	2.6±0.3

Source: Republished with the permission of John Wiley & Sons from Ref. [2]; permission conveyed through Copyright Clearance Center, Inc.

Keyence Confocal Optical Microscope (VHX-5000) was employed to measure the thickness of the islands [2]. The deposition rate on graphenated Cu increases as the electrolyte interaction time increases. However, this deposition rate begins to saturate after 6 h of dipping.

In modern VLSI technology, the typical range of interconnects is around 50 nm, which would take around 24 min to develop using the proposed dipping/treatment method with electrolyte.

Figure 5.8 shows the XRD pattern (taken from PANalytical Empyrean X-ray Diffraction (XRD) System) for the fresh graphenated Cu sample and sample treated with electrolyte for 12 h.

Figure 5.8 provides a clear indication that pure copper peaks with the orientations of {200} and {400} are present in both samples and these peaks are denoted with the "#" symbol. However, clear evidence of cubic copper oxide peaks with the orientations of {111} and {222} is presented in the sample after the 12 h electrolyte treatment as marked by the "*" symbol. It is very intriguing that the pure copper peak with the orientation of {111} is present in the sample (after the 12 h electrolyte treatment) as marked by the "Δ" symbol. It occurs because this copper {111} orientation has the minimum lattice mismatch with graphene [28], and this orientation is also considered the most favorable orientation for copper interconnects [29]. Therefore, this method is not only suitable for the growth of copper over the graphenated copper; it also produces the Cu texture that is most viable for the applications of VLSI interconnects.

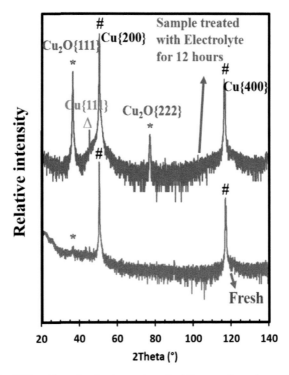

Figure 5.8 XRD pattern for fresh graphenated Cu sample and sample treated with electrolyte for 12 h. (#) Pure copper peak of {200} and {400} orientation, (*) cubic Cu$_2$O peaks of {111} and {222} orientation, and (Δ) pure copper peak of {111} orientation are observed. Republished with the permission of John Wiley & Sons from Ref. [2]; permission conveyed through Copyright Clearance Center, Inc.

5.5 Mechanism of Electroless Cu Deposition on Graphenated Cu

The aforementioned electroless Cu deposition on graphenated Cu without any additive has been demonstrated with good promises. Its underlying mechanism can be explained with the help of the energy band diagram and charge transfer.

Polarization effect is studied and shown in Fig. 5.9a with the help of density of states (DOS) [30] using the energy levels before copper (Cu)–graphene (Gr) contact with each other. When Cu and Gr interact, a dipole generates because of the polarization effect within the interface separation $d_{Cu\text{-}Gr}$ (3.26 Å in this case) as depicted in Fig. 5.9b [31]. Generally, the necessary energy (per carbon atom, ΔE_f) to isolate the graphene from the surface of copper is a positive value when the metal work function is larger than graphene and vice versa. However, it is found from the first-principles study of charge transfer by Khomyakov et al. [32] that if $d_{Cu\text{-}Gr}$ is less than 3.3 Å, the ΔE_f value is negative. Therefore, the positive and negative charges exist at the surfaces of copper and graphene, respectively. This phenomenon is also proven experimentally where graphene can act as n-doped material (ΔE_f of up to −0.3 eV) after coming into contact with copper [33–35].

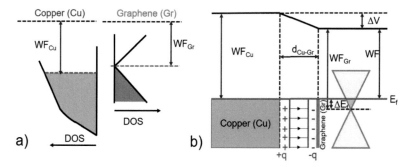

Figure 5.9 Cause of polarization effect due to the difference in the work function of copper (Cu) and graphene (Gr) depicted by (a) density of states (DOS) diagram before contact (drawn using understandings from work done by Nagashio et al. [30]). (b) Energy band diagram for graphenated copper. Here $d_{Cu\text{-}Gr}$ (3.26 Å [31]) represents the equilibrium separation of the graphene sheet from the Cu metal surface; WFCu (5.22 eV [31]), WFGr (4.40 eV [31]), and WF (4.367 eV [31]) are the work functions of copper, graphene, and graphene-covered copper (graphenated Cu), respectively; and ΔE_f (WF − WFGr) is defined as the energy (per carbon atom) required to separate the graphene sheet from the Cu surface; ΔV is the potential change generated by the metal–graphene interaction and E_f is the Fermi energy (drawn using the understandings from the work done by Giovannetti et al. [31]). Republished with the permission of John Wiley & Sons from Ref. [2]; permission conveyed through Copyright Clearance Center, Inc.

The "electric double layer" phenomena occurred due to the accumulation of ions at the graphene–electrolyte interface [36]. It happens because when graphene is immersed in an electrolyte, a difference in the electrochemical potential is created. When graphenated copper is treated with acidic electrolyte, which has plenty of hydronium ions (H$_3$O$^+$), these ions accumulate near the n-doped graphene layer at low pH level as reported [37, 38]. The band alignment of n-doped graphene and Cu^{2+}/Cu redox potential takes place under the effect of hydronium ions, which favors charge transfer in the form of electrons from graphene to the Cu^{2+} ions, as shown in Fig. 5.10, which further reduced to copper domains that are formed over the graphene. These copper domains get oxidized in air and become Cu$_2$O islands, which are observed in the SEM images [2].

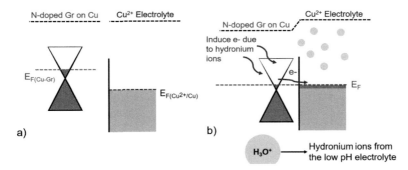

Figure 5.10 Charge transfer for electroless copper deposition using graphenated Cu as a reducing agent. Here $E_{F(Gr-Cu)}$ (~4.1 eV), $E_{F(Cu2+/Cu)}$ (5.19 eV [39]), and E_F are the energy levels of graphenated copper, Cu redox reaction, and electrolyte-treated graphenated Cu, respectively. Republished with the permission of John Wiley & Sons from Ref. [2]; permission conveyed through Copyright Clearance Center, Inc.

From the preceding discussion, it is evident that there are three important prerequisites in order to develop a metal–graphene–metal (MGM) structure by using the electroless deposition as follows:

1. ΔE_f, the energy necessary to separate the graphene sheet from the Cu surface, should be negative in graphene;
2. Low pH solution is required in order to create hydronium ions (H$_3$O$^+$) that are essential in excess amount for the energy leveling of graphene and copper;

3. The redox energy level of the metal is higher than the energy level of n-doped graphene.

This study helps to develop an understanding in order to develop MGM structures. The same analogy can also be established for other metals as well on the basis of fundamental parameters, which are equilibrium separation of the graphene sheet from the metal surface (d_{M-Gr}), ΔE_f, and a difference in the energy level of n-doped graphene and redox energy level of the metal. With such MGM, the standard interconnect back-end process can be performed with just an additional step of immersing the wafer in electrolyte for less than an hour to create the MGM structure.

The trend of ΔE_f with respect to d_{M-Gr} is depicted in Fig. 5.10 for various metals based on the work of Khomyakov et al. [32] using the first-principles study of charge transfer. It can be clearly seen that for specific d_{M-Gr} value (for various metals, it ranges from 3.2 to 3.6 Å), ΔE_f is negative for the metals that are commonly used in the VLSI technology, such as Au, Ag, and Al.

Copper can also be deposited over graphene if any of the metals (such as Ni, Co, Ti, Al, Ag, or Au) are used as the substrate at the acidic solution (i.e., low pH value) in Cu^{2+} ions environment (electrolyte) with n-doped graphene in Fig. 5.11.

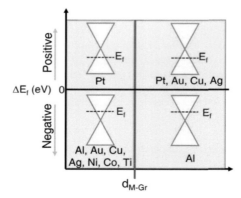

Figure 5.11 Trend of ΔE_f with respect to d_{M-Gr} for metals such as Pt, Al, Au, Ag, Cu [31, 32], and Ni, Co, Ti [30]. Republished with the permission of John Wiley & Sons from Ref. [2]; permission conveyed through Copyright Clearance Center, Inc.

5.6 Summary

As graphene layer has much better electrical conductivity, the overall resistance of interconnect for VLSI applications is much reduced. Also, graphene has excellent thermal conductivity, and thus the temperature gradient will be very small, and all these will improve the EM and conductive performance of interconnections in VLSI.

To leverage on the properties of graphene on interconnects, a newly proposed MGM interconnect structure is suggested. This could be one of the potential solutions to the interconnect challenges in VLSI as pointed out by the International Technology Roadmap for Semiconductors (ITRS) due to the continuous downscaling. In these structures, the current conduction generally occurs through the graphene, which possesses the highest electrical and thermal conductivities in addition to advantageous properties such as negligible Joule heating and temperature gradient. These structures can play a very decisive role in the current VLSI technology in terms of electromigration reliability and current conduction performances, and simultaneously it will also behave like metal–metal interconnection. The fabrication of the MGM structure using electroless plating without the use of any reducing agent is also demonstrated.

This proposed sandwiched structure is very versatile in nature as it is feasible to fabricate with all the present copper interconnect technology without making any external changes. However, the electrical conduction mechanism will be different internally as electric current will be done through graphene and help to overcome the interconnect reliability issue.

This proposed method is environmentally friendly in nature because in this electroless deposition process, no reducing agents are used, whereas the currently reducing agents are used at such a great level that they can adversely harm the environment and humans as well. These reducing agents can be toxin (such as glyoxylic acid, hypophosphite), non-biodegradable (such as EDTA), or biohazards (such as formaldehyde, hypophosphite). In contrast, it is also an economical production process as no reducing agents are used in it, as graphene acts as a reducing agent in itself in the electroless process.

References

1. C. M. Tan, *Electromigration in ULSI Interconnections*. World Scientific, 2010.

2. U. Narula, C. M. Tan, and E. S. Tok, "Metal on graphenated metal for VLSI interconnects," *Adv. Mater. Interfaces*, **5**, 13, pp. 1–10, 2018.

3. B. C. Johnson, "Electrical resistivity of copper and nickel thin-film interconnections," *J. Appl. Phys.*, **67**, 6, pp. 3018–3024, 1990.

4. J. T. Smith, A. D. Franklin, D. B. Farmer, and C. D. Dimitrakopoulos, "Reducing contact resistance in graphene devices through contact area patterning," *ACS Nano*, **7**, 4, pp. 3661–3667, 2013.

5. W. S. Leong, H. Gong, and J. T. L. Thong, "Low-contact-resistance graphene devices with nickel-etched-graphene contacts," *ACS Nano*, **8**, 1, pp. 994–1001, 2014.

6. B. Ma, C. Gong, Y. Wen, R. Chen, K. Cho, and B. Shan, "Modulation of contact resistance between metal and graphene by controlling the graphene edge, contact area, and point defects: An ab initio study," *J. Appl. Phys.*, **115**, 18, 2014.

7. C. M. Tan, W. Li, Z. Gan, and Y. Hou, *Applications of Finite Element Methods for Reliability Studies on ULSI Interconnections*. London: Springer Science & Business Media, 2011.

8. C. M. Tan, Y. Hou, and W. Li, "Revisit to the finite element modeling of electromigration for narrow interconnects," *J. Appl. Phys.*, **102**, 3, 2007.

9. D. Yoon, Y.-W. Son, and H. Cheong, "Negative thermal expansion coefficient of graphene measured by Raman spectroscopy," *Nano Lett.*, **11**, 8, pp. 3227–3231, 2011.

10. R. Faccio, P. A. Denis, H. Pardo, C. Goyenola, and Á. W. Mombrú, "Mechanical properties of graphene nanoribbons," *J. Phys. Condens. Matter*, **21**, 28, p. 285304, 2009.

11. H. Abé, K. Sasagawa, and M. Saka, "Electromigration failure of metal lines," *Int. J. Fract.*, **138**, 1–4, pp. 219–240, 2006.

12. K. Sasagawa, S. Fukushi, Y. Sun, and M. Saka, "A numerical simulation of nanostructure formation utilizing electromigration," *J. Electron. Mater.*, **38**, 10, pp. 2201–2206, 2009.

13. L. Yu, L. Guo, R. Preisser, and R. Akolkar, "Autocatalysis during electroless copper deposition using glyoxylic acid as reducing agent," *J. Electrochem. Soc.*, **160**, 12, pp. D3004–D3008, 2013.

14. S. M. El-Raghy and A. A. Abo-Salama, "The electrochemistry of electroless deposition of copper," *J. Electrochem. Soc.*, **126**, 2, p. 171, 1979.
15. S. H. Cho, S. H. Kim, J. G. Lee, and N. E. Lee, "Micro-scale metallization of high aspect-ratio Cu and Au lines on flexible polyimide substrate by electroplating using SU-8 photoresist mask," *Microelectron. Eng.*, **77**, 2, pp. 116–124, 2005.
16. S. Djokic, "Electroless deposition: History, status and future directions." In *225th ECS Meeting*, May 11–15, Orlando, FL, 2014.
17. R. Schumacher, J. J. Pesek, and O. R. Melroy, "Kinetic analysis of electroless deposition of copper," *J. Phys. Chem.*, **89**, 20, pp. 4338–4342, 1985.
18. H. Kind et al., "Patterned electroless deposition of copper by microcontact printing palladium(II) complexes on titanium-covered surfaces," *Langmuir*, **16**, 16, pp. 6367–6373, 2000.
19. L. Yu, L. Guo, R. Preisser, and R. Akolkar, "Autocatalysis during electroless copper deposition using glyoxylic acid as reducing agent," *J. Electrochem. Soc.*, **160**, 12, pp. D3004–D3008, 2013.
20. W.-T. Tseng, C.-H. Lo, and S.-C. Lee, "Electroless deposition of Cu thin films with CuCl[sub 2]-HNO[sub 3] based chemistry: I. Chemical formulation and reaction mechanisms," *J. Electrochem. Soc.*, **148**, 5, p. C327, 2001.
21. R. Touir, H. Larhzil, M. Ebntouhami, M. Cherkaoui, and E. Chassaing, "Electroless deposition of copper in acidic solutions using hypophosphite reducing agent," *J. Appl. Electrochem.*, **36**, 1, pp. 69–75, 2006.
22. J. M. Blackburn, D. P. Long, A. Cabanas, and J. J. Watkins, "Deposition of conformal copper and nickel films from supercritical carbon dioxide," *Science*, **294**, 5540, pp. 141–145, 2001.
23. H. Lee, S. M. Dellatore, W. M. Miller, and P. B. Messersmith, "Mussel-inspired surface chemistry for multifunctional coatings," *Science*, **318**, 5849, pp. 426–430, 2007.
24. Y. Sui et al., "Low temperature synthesis of Cu_2O crystals: Shape evolution and growth mechanism," *Cryst. Growth Des.*, **10**, 1, pp. 99–108, 2010.
25. P. Keil, R. Frahm, and D. Lützenkirchen-Hecht, "Native oxidation of sputter deposited polycrystalline copper thin films during short and long exposure times: Comparative investigation by specular and non-

specular grazing incidence X-ray absorption spectroscopy," *Corros. Sci.*, **52**, 4, pp. 1305–1316, 2010.

26. K. Self and W. Zhou, "Surface charge driven growth of eight-branched Cu$_2$O crystals," *Cryst. Growth Des.*, **16**, 9, pp. 5377–5384, 2016.

27. R. A. Mackay and W. Henderson, *Introduction to Modern Inorganic Chemistry*. CRC Press, 2002.

28. T. O. Terasawa and K. Saiki, "Growth of graphene on Cu by plasma enhanced chemical vapor deposition," *Carbon N. Y.*, **50**, 3, pp. 869–874, 2012.

29. KYAWTHETLATT(NUS), "Electromigration characteristics of copper interconnects," *NUS Blog*, 2014. [Online]. Available: http://blog.nus.edu.sg/kyawthetlatt/2014/08/15/electromigration-characteristisof-%0Acopper-interconnects/. [Accessed: 05-Oct-2017].

30. K. Nagashio and A. Toriumi, "Density-of-states limited contact resistance in graphene field-effect transistors," *Jpn. J. Appl. Phys.*, **50**, 7 PART 1, 2011.

31. G. Giovannetti, P. A. Khomyakov, G. Brocks, V. M. Karpan, J. Van Den Brink, and P. J. Kelly, "Doping graphene with metal contacts," *Phys. Rev. Lett.*, **101**, 2, pp. 4–7, 2008.

32. P. A. Khomyakov, G. Giovannetti, P. C. Rusu, G. Brocks, J. Van Den Brink, and P. J. Kelly, "First-principles study of the interaction and charge transfer between graphene and metals," *Phys. Rev. B Condens. Matter Mater. Phys.*, **79**, 19, pp. 1–12, 2009.

33. R. Blume et al., "The influence of intercalated oxygen on the properties of graphene on polycrystalline Cu under various environmental conditions," *Phys. Chem. Chem. Phys.*, **16**, 47, pp. 25989–26003, 2014.

34. A. L. Walter et al., "Electronic structure of graphene on single-crystal copper substrates," *Phys. Rev. B Condens. Matter Mater. Phys.*, **84**, 19, pp. 1–7, 2011.

35. A. Dahal, R. Addou, H. Coy-Diaz, J. Lallo, and M. Batzill, "Charge doping of graphene in metal/graphene/dielectric sandwich structures evaluated by C-1s core level photoemission spectroscopy," *APL Mater.*, **1**, 4, 2013.

36. C. Mackin et al., "A current–voltage model for graphene electrolyte-gated field-effect transistors," *IEEE Trans. Electron Devices*, **61**, 12, pp. 3971–3977, 2014.

37. P. K. Ang, W. Chen, A. T. S. Wee, and K. P. Loh, "Solution-gated epitaxial graphene as pH sensor," *J. Am. Chem. Soc.*, **130**, 44, pp. 14392–14393, 2008.

38. D. J. Cole, P. K. Ang, and K. P. Loh, "Ion adsorption at the graphene/electrolyte interface," *J. Phys. Chem. Lett.*, **2**, 14, pp. 1799–1803, 2011.
39. R. E. Novak, J. Ruzyłło, and Electrochemical Society. Electronics Division., *Proceedings of the Fourth International Symposium on Cleaning Technology in Semiconductor Device Manufacturing*. Electrochemical Society, 1996.

Index

activator 50
additives 91, 100
AFD *see* atomic flux divergence
amorphous carbon 27–29, 31, 33–39, 41, 45–50, 55–57, 60, 61
amorphous carbon layer 29, 40–42, 44, 49, 56, 60
analysis of variance statistics (ANOVA statistics) 77–79
annealing 29, 40–42, 54, 71, 72, 83, 85, 92, 93, 95
annealing temperature 41, 45, 57, 60, 73, 80, 85, 88
annealing time 35, 41, 43–46, 50–54, 56, 57, 70, 72, 73, 80, 83, 84, 88
ANOVA statistics *see* analysis of variance statistics
atomic flux divergence (AFD) 97–99

back-end-of-line process (BEOL process) 8, 22
ballistic conduction 4
BEOL process *see* back-end-of-line process
Black's equation 9
Brillouin zone 32, 33
Brodie's oxidation method 19

carbon 27, 35, 40, 41, 51, 53, 56, 58
carbon atom 9, 41, 51, 52, 56, 104
carbon radical 51–54, 56, 58, 61
carbon source 22, 27–29, 36, 60
catalyst 28, 45, 60, 61
cavitation phenomenon 18

chemical vapor decomposition (CVD) 19, 22, 28, 60
compressive stress 53, 56, 57
contact resistance 5, 8
copper 2–6, 8–10, 28, 29, 34, 36, 41, 42, 45, 47–50, 52–54, 56, 58, 60, 61, 91, 92, 94–96, 100–106
  electroless 100
  pure 95
  sputtered 60
copper film 22, 40, 41, 49–52, 54, 57, 58, 60
copper interconnect 2, 100–102
  graphenated 91
cracks 27, 49
crystallization 35, 44, 50, 51
current density 1, 3, 5, 6, 8, 9, 11, 96
CVD *see* chemical vapor decomposition

defect density 20, 43, 75, 84, 86, 88, 93
defective domains 93
defective sites 52, 53
defects 5, 8, 27, 33, 34, 44, 52, 80, 86
degradation process 4
deposition 4, 18–20, 100
  electroless metal 91
  island 101
design for reliability 68
design of experiment (DoE) 65–71, 73–77, 79, 81, 83–87
device 6, 10, 11
diffusion 3, 52–54, 56, 58, 60

Index

DoE *see* design of experiment

electrical conductivity 21, 22, 107
electroless deposition 100, 105, 107
electrolyte 101–103, 105, 106
electrolyte treatment 101, 102
electromigration 2, 91
environment 20, 29, 107
   aqueous 101
   cryogenic 5
   pure hydrogen 60
etching 17, 51–53, 56, 57
etching reaction 51, 57
exfoliation method 20, 27
experiment 44, 49–51, 57, 59, 61, 66–68, 74, 88, 91, 101
   full-factorial 70
   long-term 66
   one-factor-at-a-time 68

fabrication 8, 23, 37, 60, 99, 107
   integrated circuit 100, 101
Fermi energy 104
finite element analysis 46
Fisher's method 66
fullerenes 27

glyoxylic acid 107
GNR *see* graphene nanoribbon
grain boundaries 49, 61
graphenated copper 91–97, 99–105
graphene 4–11, 17–20, 22, 27, 28, 32–37, 40, 41, 43–46, 49–53, 58, 60, 61, 72–75, 84, 86, 88, 91–93, 95, 96, 98–102, 104–107
   defect-free 18, 85
   few-layer 80, 86
   many-layer 80
   many/multilayer 84

   monolayer 18, 33
   multilayered 44
   n-doped 105, 106
   pristine 6
   self-supporting 17
   single-/few-layer 84
   unstrained 59
graphene crystallization 50, 52, 56, 60
graphene-edge connections 93
graphene film 34, 95, 96
graphene growth 27, 45, 50, 51, 65, 70, 72, 73, 75, 83, 84, 86
graphene growth process 22, 28, 47, 70
graphene layer 17, 18, 28, 33, 40, 41, 43, 44, 50–53, 56–58, 60, 80, 83, 86, 105, 107
graphene nanoribbon (GNR) 4, 5, 10
graphene sheet 104–106
graphene synthesis 17, 18, 20, 22, 23, 27, 28, 36, 51, 52, 60, 67, 75, 86, 87
graphite 17–19, 27

Hofmann method 19
Hummers method 19
hydrogen 29, 35, 50, 51, 54, 58
hydrogen gas 50, 58, 60
hypophosphite 107

interaction plot 72, 74, 76, 77
interconnection 1, 2, 4, 11, 22, 91, 100, 107
interconnect 1–4, 6, 8, 11, 22, 28, 91, 92, 95, 98, 100, 102, 107
   VLSI 20, 22, 103
interconnect structure 95, 98, 107
ion 101, 105
island 101, 102
iTO phonon 33

lateral force method 18
logic circuit 10
low noise amplifier 11

material 2, 10, 11, 19, 20, 66, 97
  antibacterial 10
  brittle 47
  electrode 10
  n-doped 104
  rare earth 67
  raw 68
  sorbent 10
  transparent 10
  zero-bandgap 10
metal 4, 20, 91, 104–107
miniaturization 2
modeling 41, 46, 91, 96, 97
multifarious process 68

nanocarriers 10, 67
N-methyl-pyrrolidone 18
N-dimethyl-formamide 18
noise immunity 4

oxidation 19, 22, 101

parameters 19, 41, 66, 70, 73–75,
    83–85, 88, 97, 98, 106
photocatalyst 11
physical vapor decomposition
    (PVD) 19, 28, 30–32, 65, 91
physical vapor deposition 28, 65,
    91
Poisson's ratio 47, 97
post-PVD annealing 29, 33, 70, 72,
    73, 84, 86, 93
PVD *see* physical vapor
    decomposition
PVD method 20, 28, 36, 41, 65, 92

quantitative DoE analysis 65,
    75–77, 84, 86

radio frequency (RF) 11, 29, 41
Raman characteristics 38, 60, 74,
    75, 93, 94, 101
Raman characterization 32, 41, 76
Raman frequency 59
Raman signature 74, 88
Raman spectrum 21, 34, 35, 37,
    38, 41–43, 45, 51, 71, 72
recipe 29, 37, 66, 68, 74, 84–86, 88
reduced graphene oxide (rGO)
    18–20
reducing agent 91, 100, 105, 107
reduction 53, 58, 60, 96, 101
  chemical 19
  photocatalyst 19
  solvothermal/hydrothermal 19
  thermal 19
reliability 2–4, 11, 92
resistance 1, 4, 99, 107
  frequency-dependent 5
  graphene wire's 7
responses 72, 73, 77, 83, 93
RF *see* radio frequency
rGO *see* reduced graphene oxide

sample 19, 21, 29, 36–38, 40, 41,
    44–48, 54, 56, 59–61, 70, 72,
    73, 83, 85, 86, 92, 93, 101–103
  annealed 32, 94
  experimental 41, 84, 95
  failed 3, 4
  fresh 101
  test 20, 70
  thin film 92
secondary ion mass spectrometry
    (SIMS) 39, 40
semiconductor industry 1, 2, 22
SIMS *see* secondary ion mass
    spectrometry
SIMS data 40, 41, 60
simulation 41, 56, 96–99
simulation model 47, 49
software industry 69

sonication method 20, 27
statistical approach 65, 66, 68, 70, 72, 74, 76, 78, 80, 82, 84, 86, 88
Staudenmaier method 19
Stoney's equation 47, 49
strain 50, 59
stress 6, 8, 47, 49, 54, 56, 58
substrate 17, 19, 20, 28, 32, 36, 42, 46–48, 92, 106

temperature gradient 96, 107
tensile stress 49, 50, 52–54, 57, 60
thermal coefficient 47, 97
thermal conductivity 5, 22, 97, 107
thermomechanical stress 46, 47, 60
thin film 28, 29, 36, 40, 49, 50, 53, 54, 58, 60, 92–95
transistor 1

ultracapacitor 10
ultra-high vacuum 19
ultrasonic waves 18

vacuum 19, 20, 59
van der Waals force 17, 18
vapor deposition process 19
variance 77, 99
very large-scale integration (VLSI) 1, 2, 20, 28, 91, 99, 107
VLSI *see* very large-scale integration
von Mises stress 48, 49, 57, 58

wavelength-division multiplex 5

X-ray diffraction (XRD) 36, 102
XRD *see* X-ray diffraction

Young's modulus 47, 97